# HEARTWARMING

# HEARTWARMING

*How Our Inner Thermostat
Made Us Human*

## Hans Rocha IJzerman

**W. W. NORTON & COMPANY**

*Independent Publishers Since 1923*

p. 199: Holt-Lunstad Table used by permission of Julianne Holt-Lunstad.

Copyright © 2021 by Hans Rocha IJzerman

For information about permission to reproduce selections from this book, write to Permissions, W. W. Norton & Company, Inc., 500 Fifth Avenue, New York, NY 10110

For information about special discounts for bulk purchases, please contact W. W. Norton Special Sales at specialsales@wwnorton.com or 800-233-4830

Manufacturing by Lake Book Manufacturing
Book design by Lovedog Studio
Production manager: Julia Druskin

Library of Congress Cataloging-in-Publication Data

Names: IJzerman, Hans Rocha, author.
Title: Heartwarming : how our inner thermostat made us human / Hans Rocha IJzerman.
Description: First edition. | New York, NY : W. W. Norton & Company, [2021] | Includes bibliographical references and index.
Identifiers: LCCN 2020025838 | ISBN 9781324002529 (hardcover) | ISBN 9781324002536 (epub)
Subjects: LCSH: Body temperature. | Body temperature—Regulation. | Body temperature—Social aspects. | Body temperature—Psychological aspects.
Classification: LCC QP135 .I39 2021 | DDC 612/.01426—dc23
LC record available at https://lccn.loc.gov/2020025838

W. W. Norton & Company, Inc., 500 Fifth Avenue, New York, N.Y. 10110
    www.wwnorton.com

W. W. Norton & Company Ltd., 15 Carlisle Street, London W1D 3BS

1 2 3 4 5 6 7 8 9 0

*To my source of warmth, my wife, Dani, and to our little kleptotherm, Julie*

# *Contents*

# HEARTWARMING

## CHAPTER 1

# Hot Beverages, Electric Blankets, and Loneliness

## *Temperature and Relationships*

Sheldon Cooper walks into the living room of his apartment and notices his two best friends, Leonard and Howard, and the tense gloom that surrounds them.

"What's going on?" he asks.

"Oh, Howard's gonna sleep here tonight. He had a fight with his mother," Leonard explains.

"Did you offer him a hot beverage?"

Leonard just stares back at Sheldon in confusion. Howard, slumped on the couch, says nothing, either.

"Leonard! Social protocol states: when a friend is upset, you offer them a hot beverage, such as tea."

"Tea does sound nice," Howard admits.

Sheldon, one of the lead characters on the television sitcom *The Big Bang Theory*, may have given rise to cynical internet memes focused on his propensity to cheer up his friends with "hot beverages," but he is certainly not alone in thinking that physical warmth and moral support go well together.[1] For centuries, songwriters and poets have linked love and caring with pleasantly elevated temperatures, while loneliness and betrayal were deemed chilly. Barbra Streisand sang about being "home again and warm all over." The

Brazilian band Jota Quest, meanwhile, belted out that "love is the warmth that heats the soul," and the Beatles claimed that happiness is "a warm gun," ironically twisting *Peanuts* cartoonist Charles M. Schulz's famous "Happiness is a warm puppy."

Even our everyday language teems with such metaphors. A "warm" person is a caring and responsive person. We may get a "warm reception" or an "icy stare." A Pole may *mówi ciepło* ("speak warmly") about another, while in France people sometimes *battre froid a quelqu'un* (literally, "beat cold to someone," which in idiomatic English would be "give someone the cold shoulder").

Back in 1946, Solomon Asch, one of the fathers of modern social psychology, found in his experiments that adding the word *warm* or *cold* to a description of a person significantly changes how others view that individual. You may be viewed as intelligent, skillful, or determined, yet that won't matter as much as whether you are also warm or perhaps cold. One who is warm, Asch discovered, is perceived as generous, sociable, good natured. To be cold is not only to lack such qualities but also to be seen as displaying their opposites: to be stingy, aloof, mean spirited.[2] Asch believed that the warm-cold dimension was fundamental to social perception. Yet it took years before science began unraveling the fact that this fundamental quality was not a product of simple linguistics or contrived metaphor. We were actually—literally—feeling the physical warmth or chill of our relationships.

Fast-forward to the twenty-first century. It's 2008, and in the majestic buildings of Yale University a rather simple experiment is being conducted. A volunteer, an undergrad student, enters the lobby of the psychology building. She is met there by a female research assistant who offers to accompany her to the fourth-floor lab, where the student is to volunteer in an experiment. The assistant's hands are full: she is carrying a cup of coffee, a clipboard, and two textbooks. The two women head for an elevator.

Inside, the research assistant asks the student to briefly hold her cup as she scribbles something on the clipboard. Soon the doors ping open, and they both step out.

Unbeknownst to the student, the first part of the experiment is already over. Once inside the lab, she will be asked to read a description of a fictional "person A," who is intelligent, skillful, industrious, determined, practical, and cautious. The student's task will be to rate A on 10 personality traits, 5 of them semantically related to either "warm" or "cold."

What the 41 undergraduate students who have participated in this particular experiment didn't know was that the researchers had already divided them into two groups. In the elevator, half of them were asked to hold a disposable cup of steaming coffee from Willoughby's, their local coffee shop. The other half were handed a cup of iced coffee. This little trick was enough to affect the students' perceptions of A. Those who had handled the hot drink thought A was significantly "warmer" than did those who had held the iced coffee. For psychologists, these findings were groundbreaking. They meant that experiencing physical warmth could actually prompt an increased judgment of psychological or social warmth.[3]

The experiment opened the floodgates of research—including my own—into the links between temperature and sociability.

If briefly holding a warm beverage was enough to make us think of others as sociable and trustworthy, could it also make us feel closer to them? Not physically closer, but psychologically and socially—in the way we talk about "close friends" or "close family members"? I became determined to find out.

A year after the coffee-in-the-elevator study, I published my own variation, together with my academic advisor at Utrecht University. We had devised a tea-in-the-lab study, in which we asked the participants to hold a cup for an experimenter while she busied herself with installing a questionnaire on a laptop computer. Once again, half of the volunteers were handed a warm container, while the rest ended up with a cold one. (In the Netherlands a few years back, iced coffee was not a common drink. Fearing that Dutch people would find the beverage weird, we substituted tea, which was much more familiar to the Dutch in both hot and cold versions.) Whether in an elevator or a lab, giving people a warm or cold drink to hold influenced the subjects' perception of others.

Now, here was our next step. We asked our subjects to view a basic evaluative scale consisting of several simple Venn diagrams drawn on a sheet of paper.[4] Each diagram consisted of two circles. At the left end of the scale the two circles barely touched one another. At the right end, the circles overlapped, nearly completely superimposed on each other. In between these extremes, the diagrams showed circles with increasing areas of overlap. We invited each student subject to assume that one of the circles represented him- or herself. The other circle, we said, represented the experimenter. What we wanted to find out was whether the drawn circles overlapped, and if they did, was the overlap almost complete or did the two circles barely touch each other? We already knew that people in better—more committed, devoted, and successful—relationships typically drew their circles as more overlapping. In our experiment, people who had held the warm cup drew circles that overlapped more than those drawn by people who had handled the cold drink. We concluded that the warm-cup people felt that their *selves* were more merged with the *self* of our experimenter. Put more simply, they felt closer to the experimenter for no other reason than that she had given them a hot beverage, not to drink but merely to hold.

As we continued with related experiments, we saw that our volunteers even started using more words indicating that they were feeling close to others. This study went something like this: Instead of giving them a warm or cold cup to hold, we placed people in rooms at Utrecht University that were either warm (72–75 degrees Fahrenheit; 22–24 degrees Celsius) or cold (57°F–64°F; 14°C–18°C). They then watched a video of white and red chess pieces moving about through space. As we asked participants to describe what they saw, a "warm" participant would tell us something like this: "I saw that the red figure was walking after the other figures, and she eliminated them. First, she grabbed the second piece on the left and then one on the right. She then moved to the back, grabbed the other figures, and moved to the front where she eliminated the other chess piece." A "cold" participant, on the other hand, said that "the pawn went on an adventure with the queen, but the queen did not like him and fled. This did no good to the family De Wit ["The White," in English], and her actions caused strife and problems. The pawn was just a jackass and let her disappear until no one was happy anymore, not even the proud king or the pawn." Our participants tended to anthropomorphize regardless of whether they were "cold" or "warm." Our "warm" participants, however, tended to rely more on verbs to describe what they saw, while our "cold" participants favored adjectives.[5]

## Mere metaphor or physiological imperative?

Thanks in part to language and to what the character Sheldon Cooper identifies as "social protocol," we are tempted to explain the association between physical warmth or coldness and social warmth or coldness dismissively, in ways that arrest rather than promote investigation and thought. Even a developmental psychologist or an educated layperson familiar with basic developmental assumptions may see in the effects we've observed an "obvious" or "self-evident" explanation. As infants, we learn the association between temperature and love when our parents care for us. These associations are reinforced through life as we repeatedly experience psychological and physiological warmth at the same time. Think of a newborn cuddled in her mother's arms, well fed and safe, protected from the cold. Such an association readily slips into our language and the metaphors we use. The association explains why we call caring people "warmhearted" and unfriendly ones "cold as ice." Later on, touching something warm, even as prosaic as a cup of steaming coffee, evokes intellectual and emotional associations with trust, inclusion, and love. Wrapping palm and fingers around a warm cup is like feeling the touch of a caring parent.

Part of the appeal of this association-based explanation is that it makes good common sense. Now, common sense is a valuable asset in much of life. We cannot afford to investigate and contemplate everything we encounter, agonizing over each decision and then struggling to justify it. Common sense is a heuristic, which enables us to learn enough or assume enough about most things that we can get on with our lives. Looking both ways before crossing a busy street is common sense, and in this case, we need no more than this heuristic. There is no value in standing on the corner mentally calculating our odds of surviving the crossing. Just look both ways.

But science is not in such a hurry. "Common sense," Albert Einstein famously said, "is the collection of prejudices acquired by the age of eighteen."[6] Science does not dismiss common sense but looks beyond it, beneath it, and beside it. In the case of links between physical and psychological temperatures, new experimental data kept pouring into journals. Both its volume and content suggested that early learning and metaphors were not sufficient to explain all those links.

In one study, conducted at the University of Toronto, 52 undergraduates were asked to play a computer game called Cyberball[7]—something that we psychologists like to use to make people participating in experiments feel excluded. Cyberball works like this. You are told you are going to play an online game in which you must catch and toss a virtual ball with two other participants. They are hidden somewhere in front of their own computers. You don't know them, and they don't know you. The game is simple. There are no *World of Warcraft* visual effects here, just crudely sketched little figures throwing a ball around. What we don't tell you is that the "others" don't actually exist. They are part of the software program, and their sole purpose is to make you feel either included or left out. If you are to be rejected, "they" will toss the ball to you once or twice and then forget your existence, playing only with each other, leaving you to stare gloomily at your screen. If the goal of the experimenters is to make you feel included, however, the other players will regularly throw the ball your way throughout the game.

In the University of Toronto experiment, after some of the volunteers had been excluded in Cyberball, they were told it was time for another, supposedly unrelated study. (Here's my tip: never trust a psychologist who tells you the "other" study is "unrelated.") In this study, the volunteers were asked to rate food products on a 7-point scale in terms of their desirability, from "extremely undesirable" to "extremely desirable." The goodies on the list were hot coffee, hot soup, apples, crackers, and Coke. The list was compiled to include some warm foods and some cold foods, both salty and sweet.

Analysis of the results revealed a clear pattern. People who were ignored in Cyberball and therefore felt rejected fancied hot foods more than did those who had received the ball more often and therefore felt included. Yet there were no differences between the rejected and the included volunteers when it came to how much they desired control foods such as Coke or apples. So, the next time you feel distant from your partner and find yourself craving hot tea or soup, consider that your hunger—in this circumstance—may have more to do with thermoregulation than with digestion. A warm hug would satisfy your craving just as well, if not better.

And if after a fight with a friend you feel like turning up the thermostat, that's normal, too. In another, related, experiment by the same University of Toronto team, rejected Cyberball players were asked to estimate the temperature of the room in which they were seated. (The information was allegedly requested by the maintenance staff.) Even though some said it was a chilly 53°F (11°C), and others guessed closer to 104°F (40°C), the mean estimate by the socially excluded players was lower than the mean estimate by those who had been included during the ball-tossing game. The difference between these means was nearly 5 Fahrenheit degrees. Rejection made the students feel physically cold.[8]

To some critics, the results of the Toronto studies seemed too good to be true. Incidentally, I was conducting my own experiments on the links between social and physical temperatures at the same time as they were, but I did not feel ready to release them just yet.

My results, when we became more confident of them, were consistent with what the Toronto researchers found. If people are made to feel different from others, I discovered, they judge air temperature to be lower. If, however, they are made to feel very much alike, they judge the temperature to be higher. And if they *read* about "warm" people—people who are loyal, friendly, helpful—the room in which they sit feels warmer, too.

In the coastal city of Sopot, Poland, my colleagues and I recruited 80 students and asked them to read a short story. Some were given a story about a man named Mark, others about a woman named Marta. In some versions, both Mark and Marta were described as caring, sensitive, loyal, and friendly. They were what you might call "warm," although we purposely did not use this word. In other versions, Mark and Marta were presented as competent, creative, precise, and efficient—all positive descriptors, but none of them implying "warmth." After the students read the stories, we asked them to evaluate the temperature of the room in which they were seated, explaining to them that the room had recently been renovated and the school needed feedback. Those who had read that Marta or Mark was caring and loyal experienced the air temperature as more than 3 Fahrenheit degrees higher than those who had learned that Marta and Mark were competent and effective. It was 69.7°F (21°C) versus 66.1°F (19°C).[9] Want to save some money on your heating bills? Perhaps you should look for a "warm" roommate.

Experiments like the one concerning Mark and Marta complicate the commonsense explanation of the links between temperature and sociability. According to the earlier, commonsense theories, these experiments should have demonstrated no links at all because metaphors act only in one direction. Physical temperature, like that of a warm cup, can activate a metaphor such as "warm personality" and make us think about trust and love. But according to the metaphor theory, thinking about trust and love should not make us perceive physical temperature as higher. Metaphors don't work this way. They can turn the concrete into the abstract but not the other way around.[10] Something else was obviously going on. But what?

We needed more clues. My colleagues and I set out to conduct another experiment. It went like this: The student participants were seated in tiny cubicles in front of rather dated computer screens on which Cyberball was running. Each student's dominant hand, the right one in most cases, rested on a computer mouse, while the index finger of the other hand was connected to a cable coiling into a sensor that measured skin temperature. In a classic Cyberball scenario, each volunteer was either made to feel included in the game or relegated to the sidelines.

After analyzing the data, an unmistakable pattern emerged. Socially rejected people ended up with colder fingers. Their skin temperature dropped, on average, 0.68 Fahrenheit degrees. And that wasn't the end of it. In a subsequent study, we offered hot tea to the excluded subjects. After simply holding the cup for 30 seconds, they felt better, reporting themselves less sad and less tense after they had warmed up their chilly fingers.[11]

As it turns out, Sheldon Cooper was onto something, but it was far more than a matter of "social protocol." The Toronto studies and our own revealed that the effects of temperature on our sociality are bidirectional. Physical temperature affects the perception of social warmth or coldness. Thinking about social warmth or coldness affects physical body temperature. This bidirectionality was our first hint that social thermoregulation is about far more than metaphors. The links between the perception of social warmth/coldness and physical warmth/coldness must be at least to some extent biological.

## Turning up the heat on development and evolution

Some hormones produced by our bodies, such as oxytocin and serotonin, are involved in both how we become attached to our parents early in life and how we regulate body temperature. Oxytocin, often narrowly and misleadingly viewed as the "cuddle hormone," is released by maternal touch. But in other species it also plays a role in physical warmth; elevated temperature prompts the release of oxytocin.[12] Mice genetically engineered to lack oxytocin receptors have trouble regulating their body temperature.[13] Serotonin, often thought to be chiefly related to feeling good, (sometimes) also relates to people being friendlier, "warmer" in the social sense.[14] The hormone's production by the body is affected by physical temperature. Studies show that rats kept in hot surroundings develop more serotonin-producing neurons in the brain stem.[15] Moreover, studies using functional magnetic resonance imaging (fMRI), which measures brain activity by detecting changes associated with blood flow, reveal that there is a considerable overlap in multiple brain areas associated with social behavior and with the regulation of temperature.[16]

As new studies kept pouring in, the picture became clearer to me: the involvement of such biological mechanisms as the oxytocin and the serotonin systems means that we have likely evolved so that social warmth relies on physical warmth. For young, helpless animals, such as human newborns, there is no way to stay warm other than being taken care of by parents, so it should be no surprise that the very basic neural mechanisms regulating temperature also govern our social relations. This means that the effects we've seen in our studies are not just the result of linguistic metaphor or even Sheldon Cooper's "social protocol." Oxytocin is not just a "cuddle hormone," and serotonin is not just the "feel-good hormone."[17] Both are critical to regulating our metabolic or energy-related resources, especially those related to warmth.

With all these biological systems at work, an animal saves energy if it can rely on huddling, thereby conserving body heat by reducing its loss to the surrounding environment. This puts it in the position of predicting its future body temperature based on its *social* capital. A penguin knows it won't freeze or starve to death burning its precious fat reserves if it surrounds itself with many reliable others. From an evolutionary perspective, those individual penguins who were bad at predicting their social capital were less likely to survive to reproduce and pass on their genes. In the game of natural selection, they were the losers—to the benefit of the gene pool.

Humans effectively build on top of this old-school, penguin-like biological foundation. They pile upon it more abstract, more "social" notions, such as trust, friendship, and love. One result of this human biological evolution was the linguistic evolution of the word *warmth* as a metaphor for such social concepts as trust, friendship, and love, which in fact are biologically linked to physical warmth. We humans are such intensely social creatures, and society is defined to so great an extent through language, that we have collectively forgotten the original links between physical temperature and the social concepts of trust, friendship, and love and today recognize only their metaphorical links to *warmth* and *coldness*. The metaphor became nothing more than a handy expression. But as with the penguins, our brains still act as machines that combine the prediction of weather with the prediction of social capital. Whether or not we are aware of doing so, we continually evaluate social cues to tell our bodies how warm or cold we are likely going to be in the near future.

The problem is, we are not literally penguins. In the complex lives of twenty-first-century humans, relying on such archaic links between social and physical temperatures can have quite serious consequences for our relationships, a situation that is the more worrying if we are unaware of what may be guiding us.

## In the heat (or cold) of the moment

On the evening of April 19, 1989, a young woman named Trisha Meili went jogging in New York City's Central Park. A few hours later, she was found unconscious, having been brutally raped and beaten, her body chilled to 80°F. Doctors at Metropolitan Hospital in Harlem, to which she was taken, were shocked to discover that Meili had lost 75 percent of her blood. It took her 12 days to regain consciousness after the assault, most of her memories lost. The police soon arrested five young men from Harlem, four of them black and one Latino. They were charged with rape and attempted murder. The trials of what the press now called the "Central Park Five" began in August 1990.

"It was . . . the cold courtroom where it was heard that set the tone," according to the *New York Times* account of the trial of three of the defendants, Yusef Salaam and Antron McCray, both age 15, and Raymond Santana, age 14. Indeed, the section of the article that set the stage was titled "Cold, Cold Court."[18] After 10 days of deliberations, the jury handed up its verdict. The young men were found guilty of rape, assault, robbery, and riot. The two other defendants, Kevin Richardson and Korey Wise, were tried and convicted separately in December. The five served between 6 and 13 years in prison. But that wasn't the end of the story. In 2001, Matias Reyes, already in prison for murder, confessed that he had attacked the jogger—alone. DNA evidence proved definitive in corroborating his confession. The five teens had been wrongly convicted and sentenced.

In the trial of three of them, the "cold, cold court" merited prominent mention from the reporter. Did this influence the verdicts and sentences? It is impossible to determine, and it is far more likely that racial bias was more important than ambient temperature in this case. Nevertheless, research shows that the influence of room temperature cannot be ruled out and, in fact, was likely not slight, let alone negligible.

In a 2014 experiment conducted in Germany, 133 undergraduate volunteers were presented with eight mug shots and asked to guess what crime each of the arrested had committed. Unknown to the students, the lab in which they were seated was either heated to 79°F (26.1°C), kept at a mild 74°F (23.3°C), or chilled to 67°F (19.4°C). The results showed that those assigned to a cold "courtroom" were more likely to judge the criminals harshly. They attributed to the eight "criminals" crimes that would have resulted in longer prison sentences than the crimes that were guessed by the volunteers in a warm room. In the chilly surroundings, kidnapping and murder predominated; in the balmier ambience, it was drug possession and tax evasion.[19]

At the very least, if you ever find yourself facing a jury, you might hope for a well-heated courtroom—something the first three Central Park Five defendants apparently did not have. But we must admit that we do not yet have a truly convincing explanation of the "courtroom" temperature effects.

Still, in other studies from Germany, male participants were asked to look at and compare themselves with others on a warmer or colder day. In one study, the male participants viewed an obviously virile naked torso, while others looked at the torso of a markedly weaker man. If they were asked how many push-ups they could do or how long they could hold a liter of beer with their arm extended, these male participants tended to rate themselves more like the standard to which they had just compared themselves if it was warmer (no matter which standard it was). If they had seen the manly man, the male participants could act like Arnold Schwarzenegger in his heyday. If they had seen the physically weaker man, so they tended to see themselves. These effects did not emerge on colder days.[20]

Thus, under warmer conditions, people tend to assimilate to strangers. In colder conditions, on the other hand, people may think more about those nearer and dearer to their hearts. As a result, they may take strangers less into their considerations. Taken together, these results hint at why people might harbor negative thoughts about criminal defendants while in chilly surroundings, contrasting themselves with the criminal: "He is not like me." Yet, in that same cold environment, they might think about the cuddly "warm" love of their loved ones. This, in turn, may make the thought of crime and threat even more menacing and appalling.

## The impact of attachment style

Whether we are judging someone's personality in terms of social warmth or cold, or someone else's fate in terms of legal guilt or innocence, temperature influences our "verdict"; this is also linked to how we form relationships with others. Attachment theory attempts to describe the dynamics of interpersonal relationships, and most psychologists believe that we develop different types of "attachment style" in childhood. The great developmental psychologists John Bowlby and Mary Ainsworth identified three major styles: secure attachment (which develops when children feel they can rely on their parents), anxious-ambivalent attachment (which is associated with inconsistent parental responses), and anxious-avoidant attachment (which is associated with unresponsive maternal behavior), later supplemented by disorganized attachment (a combination of ambivalent and avoidant attachment).[21] One of the most crucial aspects that a parent must attend to is whether the child feels too cold or too warm.

My own studies have also revealed that the effects of temperature on our behavior are linked to our attachment style. Here is an experiment I once conducted with my colleagues. We invited 60 kindergarten children to a classroom in a school in Abcoude, a small Dutch town a stone's throw south of Amsterdam. First, the children were tested for their friendship attachment styles. We asked them such questions as "Do you find it easy to become good friends with other children?" and "How do you feel without good friends?" We wanted to be certain that the little ones understood what it was that we wanted from them. And to help them understand how we used scales to answer our questions, my student Emma Landstra created additional questions so that the children could practice with a topic about which they had very strong opinions, such as "How much do you like brussels sprouts?" They answered on a numerical scale, but with arrows so that it was clear for the children how to answer (see figure).

| ← | | | ? | | → | |
|---|---|---|---|---|---|---|
| Not at all | No | Not really | I don't know | A little | Yes | Very much |
| O1 | O2 | O3 | O4 | O5 | O6 | O7 |

Then it was time to play a little game. Some of the children were led into a chilly room, where the thermostat was set between 59°F and 66°F (15°C and 19°C). Others were shown to a cozy room with air temperatures of about 70°F–78°F (21°C–26°C). We gave each child 10 equal-sized stickers or 20 balloons, and asked how many they would give to another, unknown child "next door," who, we told them, would not be getting any gifts at all. The kid next door didn't really exist, but our young volunteers were not aware of that.

When we analyzed all the data, we discovered that in a warm room, securely attached children gave more stickers to the kid next door, almost three on average, than did securely attached children in a chilly room, who gave about one and a half on average. For insecurely attached children, room temperature did not really matter. In both the warm-room and cold-room scenarios, the number of trinkets they shared was essentially the same—about one.[22]

We weren't exactly surprised. Attachment and temperature are powerfully linked. For example, Mary Ainsworth discovered that mothers of insecurely attached babies show signs of aversion to close bodily contact.[23] We also know that newborns who are held skin-to-skin have relatively small differences between skin temperature and core temperature.[24] It appears that the biological links between physical temperature and psychological temperature become activated only if we have experienced early in life that physical warmth and caring for others go hand in hand—and if we are securely attached.

First, then, comes "the wiring," our genetic makeup. On top of that is the learning.[25] Young children usually learn that expressing a need prompts their parents to respond in ways that address the need. In the case of a temperature need, "Mommy, I'm cold" may elicit a cuddle, an extra blanket, or an upward dialing of the bedroom thermostat. The physiological, physical, and emotional links forged in this early learning endure throughout our lives. Whenever we experience comfortably elevated physical temperatures, concepts of social closeness, friendship, and trust are activated. On the other hand, experiencing a psychological chill such as rejection or betrayal makes us perceive ambient physical temperatures as being lower than they are, quite possibly because our skin temperature physically drops down. And being the complex animal that we are, it's probably even more complicated than that.

What, then, do we know? Our brains become "weather prediction machines," forecasting our social temperatures. Can our friends and family be trusted? Can they be counted on to warm us up in times of need? To keep us safe from dangerous chills? In the past, before the invention of central heating and triple-rib king-size electric blankets, such social-weather reports could be matters of life and death.

## The dubious "triumph" of central heating

Back in 1779, in a room on the rue de la Mortellerie, smack in the center of Paris, a manual laborer named Louis Bequet shared his one and only bed with his wife and their five children: seven people cuddled on a single mattress. The laborer's hard fate, similar to that of so many others in prerevolutionary France and, for that matter, all of Europe, was meticulously inscribed in the archives of the city of Paris. An examination of these documents reveals that very few eighteenth-century Parisians enjoyed a bed to themselves. Fewer than 10 percent of children had cradles or cots, and servants shared their beds with other servants, who were often virtual strangers. In sum, between 1695 and 1715, on average, 2.3 Parisian servants occupied one bed.

Sleeping in canned-sardine fashion did have an upside. "Warmth was created by bodily contact. Comfort was a matter of bodies," wrote French historian Daniel Roche in his *Le peuple de Paris* (available in English as *The People of Paris*). A human body radiates about 330 Btu (British thermal units) of energy per hour—roughly equivalent to the energy radiated by a classic tungsten-filament 100-watt lightbulb. Keeping a small bedroom of about 100 square feet warm during a relatively mild Parisian winter requires about 1,100 watts per hour. So, although the seven people in Louis Bequet's family bed were not enough to keep their bedroom truly warm, their radiated body heat could cut space-heating requirements by over 60 percent. And this is based only on physical heat. It does not account for the social and emotional warming effects of huddling. Nevertheless, we do know that, when packed tightly together, nonhuman animals can save up to 53 percent of the energy required to heat up their bodies. At night, the Bequet family might have been quite cozy, even without a fireplace roaring in the room.[26]

In centuries past, co-sleeping was the norm for humans. As late as the mid-nineteenth century, in the parish of Tullaghobegley, Ireland, you could find nine people to a bed, divided according to sex. That is, women would sleep in one direction, men in the other, their feet placed between their opposite numbers' heads.[27] They were warm and snug, penguin-like. The physical connection between warmth and sociability, between coziness and trust, made perfect sense. Yet, in modern times, that physical link has been largely broken, at least when we are talking about adults. The reason for this is the triumph of central heating.

That triumph was a long time coming. As far as we can tell, central heating was invented in ancient Greece and further developed in Rome. In the astoundingly well-preserved Stabian Baths at Pompeii, for example, the so-called hypocaust system, something resembling our modern floor heating, involved raising the floor about 27 to 35 inches on small pillars and, in the space between, circulating hot air from a furnace. As with so many other advances in civilization, central heating and the hypocaust system vanished with the demise of the Roman Empire and the onset of Europe's Dark Ages. Once again, human beings had to rely on the warmth of one another's bodies, huddling in shared beds. It was not until the nineteenth century that central heating began to reemerge and make its slow conquest of our houses.

Thanks to technological advances in central heating, those of us who live in well-developed nations today need the physical proximity of other human beings less than our ancestors did. Today, as adults, we don't have to scan "social-weather forecasts" to evaluate whether our close ones will be there in times of need, at least in times of thermal need. Does this make us moderns feel more independent and free? Are we less inclined to rely on others, to seek connection? Since there are no studies on how diminished huddling might have changed us and our societies, we can only speculate on the answers. Yet one thing is for sure. Our basic wiring is mostly unchanged. We still have those deeply ingrained, rat-like neural links between psychological and physiological warmth, and we still learn early in life that *snug* equals *loved*—at least most of us learn this.

By reducing the physical-survival need for huddling, central heating may pose a challenge to how we human beings use social thermoregulation to create, strengthen, and maintain social and emotional bonds. The proliferation of even-more-modern technologies, such as electronic and digital communication, has allowed us to connect with one another in the total absence of physical proximity, let alone contact. We can hear and see one another in real time despite separation by great distance. And yet this ability, remarkable though it is, serves to remind us of all that we lose when physical proximity is absent.

Touch and warmth are powerful dimensions of human communication. We feel their power most poignantly when we are deprived of them by a technology that falls short when it comes to serving these dimensions. And when technology falls short unexpectedly, when, for example, the furnace fails and we find ourselves without central heating, we may be forced to discover what it means to have to rely on others to huddle with us and provide comfort-giving and even life-giving warmth. In such extremity, even today, we can be dramatically reminded of how, throughout evolution, the reliance of social warmth on physiological warmth became hardwired into our bodies, just as it is in other animals such as penguins, naked mole rats, or the Barbary macaques of Morocco.

Our more distant ancestors needed fellow hominins for thermoregulation, as a way to reduce their individual energy needs. Those individuals who were good judges of the reliability of others, and skilled at predicting the social weather that surrounded them, survived to pass on their genes. Over millennia, more-advanced cognitive systems were scaffolded on top of that basic setup. They recycled the ancient wiring to process complex social information. As newborns today, we test our ancient circuitry. In the arms of our doting parents, most of us confirm that physical warmth does indeed equal safety and love. In later childhood and adulthood, even handling a warm beverage or a cold pack can activate these systems. That is why, in my experiments, those who were asked to hold a warm cup of coffee felt closer to experimenters than did those who held a cold cup. That is why thinking of friendly, caring people made Polish students feel physically warmer. Elevated temperature signals to us that others are close by, while, in contrast, being cold suggests we are lonely. In the modern world, an environment characterized by intensive digital connection, being excluded from full participation in an online game can make the room you're in feel cold—as cold as Louis Bequet felt in 1779 when he went off to work on a chill Paris morning, leaving his wife and five children still snug in their common bed.

In our twenty-first-century society, the social-weather reports are much more complicated than they ever were for our ancestors. We may have escaped, through a combination of mechanical and electronic means, many of the barriers of time and space, but we cannot escape or conquer or even reliably suppress every aspect of our physical environment. Physical temperature can influence decision making in courts, so that with the mercury down the accused appear more cold-blooded. More than ever, it is vital to understand how and why temperature leads us to feel and act in certain ways. Where does this all come from? The next chapter embarks on an adventure into the still-emerging field of study that flows from "embodied cognition," a theory that responds to the long-cherished Cartesian view of a strict mind-body dualism by discovering hitherto unrecognized connections between the whole body and the mental processes of cognition.

## CHAPTER 2

# The Human Machine

## Temperature and Embodied Cognition

Numerous studies support Sheldon Cooper's confidence in the social potency of hot beverages. Clearly, temperature leads us not only to feel but also to act in certain ways. What are the processes and mechanisms of these influences? The how and the why? To answer these questions, we need to dive into the history of cognitive science and its evolving models of cognition in general and of embodied cognition in particular.

We begin in the late Renaissance with René Descartes (1596–1650), one of the prime movers of the Scientific Revolution and a contributor to the fields of philosophy, mathematics, and science, including psychology. His *Les passions de l'âme* (*The Passions of the Soul*, 1649) is a pioneering treatise on emotions, about which, he promised his readers, he would write "as if no one had written on these matters before." He questioned everything, and he seemed to have an answer for everything, too. His most famous answer was a response to "radical doubt," the philosophical position that knowledge is ultimately impossible. "*Je pense, donc je suis*," he wrote. It was later translated into Latin as "*Cogito, ergo sum*," and is usually rendered in English as "I think, therefore I am."[1]

It is a pretty good answer, but, like all syllogisms, it does not dig very deeply. Descartes went deeper when he considered the relationship of the mind to the body. He posited that the mind exists inside the body like "a pilot in a ship."[2] The metaphor suggests that the mind exerts influence over the body but the body exerts no influence over thought. Descartes went even deeper than metaphor, suggesting that what he called the *epiphysis cerebri*, the pineal gland within the brain, was the seat of the soul. It, not the brain itself, was where thoughts were formed. When the soul moved the pineal gland to form a thought, spirits were driven toward the pores of the brain, which directed them via the nerves to the muscles, and thus moving the limbs was ultimately directed by the brain.

Immaterial mind, for Descartes, was no more a part of the material brain than a pilot was part of the ship he steered. Nevertheless, just as the *flesh-and-blood* pilot unquestionably influenced the course of the *wooden* ship, so the *immaterial* soul, via the pineal gland, influenced the *material* brain to move the *material* limbs of the *material* body. Though distinct from matter, mind influences matter. This is what philosophers call Cartesian dualism, and because of certain developments in our modern times, the idea of Cartesian dualism has had some very important implications for our understanding of human functioning. The relationship in this theory between mind and matter is essentially one-way: thought influences body, but body exerts no influence over thought.

Allowing for some variation, Cartesian dualism describes the twentieth-century view of cognition, which holds that our thinking is localized in our heads. As we will see, there are several reasons to conclude that this view is too limited. One of the most important of these is social thermoregulation, which could not operate if cognition were limited to the mind and not also in various ways embodied throughout the organism. So, just how should we think about thinking?

# From Descartes to Turing

If Descartes was a founding father of the Scientific Revolution, Alan Mathison Turing (1912–1954) was a founding father of the yet-ongoing Artificial Intelligence (AI) Revolution. Trained as a mathematician, Turing published in 1936 a groundbreaking theoretical paper titled "On Computable Numbers, with an Application to the *Entscheidungsproblem*," in which he imagined in detail a "universal computing machine" capable of performing any calculation that could be represented by an algorithm.[3] This thought experiment was an alternative to a formal mathematical solution to the *Entscheidungsproblem* ("decision problem") mentioned in the title of the paper. In other words, Turing can be credited with inventing—schematically, on paper—the modern computer. As anyone knows who has seen the 2014 Turing biopic *The Imitation Game*, during World War II he built an electromechanical computer capable of breaking the infamous German military "Enigma" cipher. After the war, in 1950, while teaching at the University of Manchester, he published "Computing Machinery and Intelligence," a paper in which he considered the question "Can machines think?"[4]

To begin with, Turing believed that "thinking" was a concept that defied meaningful definition. So he asked a less ambiguous question: Are there "imaginable digital computers" that could excel in the "Imitation Game"? As he explained in the paper, the Imitation Game is a parlor game for three players. Player A is a man, player B is a woman, and player C, "the interrogator," may be of either sex. Player C cannot see players A or B but can communicate with them using written notes. By asking questions of A and B, C tries to determine which one is the man and which one the woman. Player A is assigned to trick the interrogator into making the wrong decision, and player B is assigned to assist the interrogator in making the right one. The paper concludes that a digital computer playing the Imitation Game should be able to output results indistinguishable from those output by a human player. For this reason, it is impossible to argue definitively against the proposition that machines can think.

Turing's conclusion is not only fundamental to the philosophy, science, and technology of AI but is also a modern confirmation of Cartesian dualism. Turing speculated that digital computers—machines—would increasingly come to resemble human beings because, at least as evaluated by input received and output produced, there is little difference between "thinking" produced by a human brain and "thinking" produced by a machine brain. But what about the biological tissues of the body? So far as the end product is concerned, these exert no influence over thinking. Whether the "mind" in question is human or digital, the function is that of a computer calculating algorithms. It is a Turing machine.

## Descartes and Turing challenged

Cartesian dualism has never been universally accepted, and Turing, after all, did not finally argue that there is no difference between the way people and machines "think." His argument was that, based only on input and output, it is impossible to distinguish the *product* of human cognition from that of computer cognition. That being the case, it is fair to say that both machines and human beings "think." Turing described an imaginary machine capable of applying algorithms to solve problems. In so doing, he laid the foundation for the modern theory of computer science by identifying the three necessary major components of a digital computer: a means of data storage, an executive unit, and a means of control. This "Turing machine" has also become widely used as a metaphor for the human mind. Descartes speculated that the pineal gland was the seat of the soul. Unlike him, Turing did not claim to have explained the organs of the mind. The three components of a Turing machine may provide a suggestive metaphor of cognitive processes, but they are neither an anatomy nor a physiology. Indeed, for Turing, the human mind remained a black box, something into which data was admitted and from which results ("thought") emerged.

None of this means that the mind-body dualism, or Cartesian dualism, is false, but recent studies do suggest that the human mind is not just an executive unit operating independently from the body, a pilot *in* the ship but not *of* the ship. In fact, the studies present evidence that our bodies are entwined far more inextricably with our heads than Descartes or Turing inferred or imagined. For instance, wearing a heavy backpack makes slopes appear steeper—not merely *feel* steeper, but cognitively *appear* to be steeper. In studies in Charlottesville, Virginia, participants were asked to estimate the steepness of a hill while standing at its bottom. This hill appeared to be steeper when participants wore a heavier backpack or when they reported themselves as being tired.[5] But put someone on a lofty balcony, and their estimate of the distance to the ground is related to their fear of falling.[6] Or consider: we think differently about *time* when we move forward in *space* than when we stand still. Stanford University students were asked the ambiguous question "If Wednesday's meeting has been moved forward two days, what day is it now?" If they themselves had just moved forward in a fast-moving lunch line, they thought about the meeting as being moved to Wednesday. But those who had been standing still? Then the meeting, they answered, would be on Monday. If the mind has no direct connection to the body, why should scheduling our meetings require—or at least get a boost from—motion in our bodies?[7]

Once you begin to think about such links between mind and body, it is difficult to stop. You begin seeing them everywhere. Take the association between feeling guilty and feeling compelled to wash your hands. We find this apparent mind-body link in the Bible: "When [Pontius] Pilate saw that he could prevail nothing, but that rather a tumult was made, he took water, and washed his hands before the multitude, saying, I am innocent of the blood of this just person: see ye to it."[8] And, of course, we find it in Shakespeare. Having goaded her husband into killing the good King Duncan and others equally innocent, Lady Macbeth, consumed with guilt, appears to wash her hands while sleepwalking. "It is an accustomed action with her, to seem thus washing her hands: I have known her continue in this a quarter of an hour," her lady-in-waiting explains to a doctor summoned to treat her presumed madness.[9]

Scripture and literature do not provide the only descriptions linking guilty feelings with acts of ablution. Psychological studies published in 2006 and 2010 seem to demonstrate a "priming" effect (exposure to one stimulus activates our thoughts and in turn leads us to respond to a subsequent stimulus) in that the response to a cleaning cue was increased after it had been induced by a feeling of shame. For instance, in the 2006 study, subjects were asked to recall a good or bad past deed for which they were responsible. Next, they were asked to fill in the letters of three incomplete words: W_ _ H, SH_ _ ER, and S_ _ P. Subjects who had been asked to recall a bad deed were about 60 percent more likely to respond with WASH, SHOWER, and SOAP than with such noncleansing alternatives as WISH, SHAKER, and STOP.[10]

This is where most books on embodied cognition would have stopped, their authors feeling they had made their case. But the description of these effects is not enough. We need a further understanding of the mechanisms behind them. The results I have just described were certainly intriguing, but the problem was that when other researchers repeated the 2006 study with larger numbers of people participating, they were unable to replicate the originally reported results.[11] This was a harbinger of the replication crisis, a problem that has stalked psychology since 2011.

The replication crisis is at least in part the result of jumping to conclusions on the basis of insufficient understanding. In this book, I intend to avoid such premature leaps. What that means is that I often cannot deliver settled certainty, and this may upset some readers. In my defense, I can say only that it is not my job as a scientist to tell a pretty story. It is my job to tell the truth. And the hard truth is that psychology books should not even try to offer concrete advice or definitive instructions about what to do in life. Most psychologists do not yet command the statistical tools to be so prescriptive. The best we can do right now is what I aim to do here, which is to identify some general principles that leave readers smarter than they were before they opened the book—a plausible verbal theory, if you will, but not yet a formal one. Still, I subscribe to what Brazilian-born biologist Peter Medawar said in *The Limits of Science*: "Science is incomparably the most successful enterprise human beings ever engaged upon."[12] Of course, one of the great reasons for its success is the discipline science instills in those of us who practice it to express radical doubt about everything.

We will return later in this book to issues of replication, but for now it is important to note that such failures do not *disprove* a link between mind and body. They do, however, imply that the links—what psychologists call embodied cognition—are neither simple nor straightforward. Involvement in some program or project or task becomes distasteful, and you declare your intention to renounce all connection with it: "I wash my hands of this dirty business!" Symbolic ablution—washing your hands—is a powerful metaphor that makes the abstract concept of ridding yourself of guilt or some other emotional burden more concrete and therefore more vividly comprehensible. But just because a successful metaphor can be used to embody an abstract thought or emotion in a concrete image, it does not necessarily follow that the abstract thought or emotion will prompt (or *prime* someone to produce) a corresponding metaphor. In other words, recalling a past deed you find shameful will not necessarily prime you to see "S _ _ P" as SOAP instead of SHIP, SHOP, or maybe SLOP.

Yet examples of embodied cognition are nevertheless plentiful. The weight of that heavy backpack really does influence the cognitive act of estimating the pitch of a slope. Changing one's perspective of space has a demonstrable—and repeatable—effect on how we *think* about time. And, as we will see, our physical experience of temperature as processed by our inner thermostat demonstrably and repeatably influences our thinking. As with other members of the animal kingdom, our body influences cognition, which does not dwell wholly within the splendid isolation of our head.

## The cognitive revolution

No physiological or psychological mechanism works in isolation. Before we can go on with the rest of this book and meaningfully explore how thermoregulation operates as social thermoregulation, we need to understand thermoregulation in the context of cognitive science. This requires a bit of preliminary history.

Like many revolutions, the cognitive revolution started out as a counterrevolution: the approach it countered, behaviorism, was itself a revolt against psychoanalysis and other forms of so-called depth psychology. Many psychologists objected to depth psychology because predictions made according to its leading assumptions could not be tested experimentally. Behaviorism focused on cause and effect, stimulus and response, input and output, rather than on assumptions about cognitive processes that could not be adequately tested. In the late nineteenth century, the U.S. psychologist Edward Thorndike formulated the "law of effect,"[13] which stated, quite simply, that responses producing a satisfying effect in a given situation become increasingly *likely* to occur again in that situation; conversely, responses that produce a dissatisfying or discomforting effect in a given situation become increasingly *unlikely* to occur again in that situation. During the first half of the twentieth century, the U.S. psychologist John B. Watson built upon Thorndike's law of effect to create "methodological behaviorism," which differentiated between "public events"—behaviors of an individual that can be objectively observed—and "private events"—thoughts and feelings, which are not objectively observable. Such private events are, of course, the province of depth psychology, but Watson believed that because these events could not be objectively observed, they should be ignored by psychologists. The only valid objects available for scientific psychological study, Watson held, were public events: visible, measurable behavior.[14]

Behaviorists who followed Watson assumed that people regulate ("modify") their public, or overt, behavior according to the presumed consequences the behavior will produce. Operating from this assumption, the early behaviorists used such tools as positive and negative reinforcement to elicit or increase desirable behavior and positive and negative punishment to reduce or eliminate undesirable behavior. The internal processes by which reinforcements and punishments operated, the Watson-style behaviorists argued, were private events and therefore inherently unavailable to objective observation.

In the late 1930s, another U.S. psychologist, B. F. Skinner, introduced a new approach to behaviorism called "radical behaviorism," which rejected the notion of simply ignoring private events. He theorized that private events are not, in fact, entirely internal and cut off from the world, but are subject to some of the same environmental variables as are observable behaviors, so-called public events. Yet even though radical behaviorism holds that private events are not entirely private, it still focuses primarily on input and output and generally treats the process of cognition as a black box, perhaps not impossible to study but very difficult nonetheless.[15]

Though "radical," Skinner's approach was not immune to counterrevolution. At least two breakthrough publications contributed significantly to the intellectual context in which the modern study of cognition emerged. The first was *The Nature of Explanation*, published in 1943 by Kenneth Craik, a British psychologist and philosopher and the first director of the Medical Research Council's Applied Psychology Unit. In this seminal document of cognitive science, Craik argued that the mind constructs "small-scale models" of reality and uses them to anticipate events. This was an early iteration of what is today called a "mental model," an internal representation of external reality, which, some psychologists believe, plays an important role in cognition generally and in reasoning and decision making more specifically.[16]

Although Craik played a crucial role in the development of experimental cognitive psychology, we can never know just how far he himself might have taken the mental model concept. Fatally injured when a car collided with his bicycle in Cambridge on May 7, 1945, Kenneth Craik died at 31. But a later researcher, Jay Wright Forrester, explained in 1971 that the "image of the world around us, which we carry in our head, is just a model." None of us "imagines all the world, government or country." Instead, we carry in our minds "only selected concepts, and relationships between them," and we use these "to represent the real system."[17]

The second landmark publication was Alan Turing's 1950 "Computing Machinery and Intelligence," which we encountered earlier in this chapter. Turing pointed out that he was not trying to answer the question of whether computers can "think." Instead, he was asking if we could validly model human behavior such that we could not distinguish human cognitive behavior from computer "cognitive" behavior. The point was not whether computers can think but whether, in solving a cognitive problem, a computer can replace a person. Turing's answer was yes. More important for the study of human cognition—lifting the lid on the black box—"Computing Machinery and Intelligence" implied that it was quite possible to make inferences about "private events," internal mental processes, including cognition. Remember, this was 1950, long before the advent of fMRI, which images areas of brain activity in real time.[18]

The theories that followed from both Turing and Craik often explained cognition through the mind-as-computer metaphor, which was a very influential approach. In 1980, however, the U.S. philosopher John Searle published "Minds, Brains, and Programs" in the journal *Behavioral and Brain Sciences*. The centerpiece of the paper was a thought experiment called "The Chinese Room." Its hypothetical premise was that AI research succeeded in building a computer that behaved as if it understood Chinese. Feed it Chinese characters, and it would deliver output indistinguishable from that of a human Chinese speaker, thus passing the Turing test.[19]

But Searle asked: Does the computer *understand* Chinese? Or is it only *simulating* the ability to understand Chinese? An affirmative answer to the first question Searle called "strong AI." An affirmative answer to the second he labeled "weak AI."

Searle next imagined himself in a closed room with a book that has an English version of the computer's program. He is also supplied with plenty of paper, pencils, erasers, and filing cabinets. Searle could receive Chinese characters passed to him through a slot in the door, process them according to the program's instructions, and thereby produce Chinese characters as output. If the computer had passed the Turing test this way, Searle argued, then he would do so as well, in effect running the program manually. This meant that there was no significant difference between his role and that of the computer in this experiment. Both he and it would follow an algorithmic program, producing a behavior that an observer would interpret as demonstrating intelligent conversation.

The thing is, Searle confessed, "I don't speak a word of Chinese," and he therefore concluded that the computer could not understand the conversation either. Without "understanding," he argued, the computer had no "intentionality" and therefore was *not* thinking. If the machine did not think, it could not have a "mind." Searle concluded that strong AI is false, which meant that the computer did not work like the human mind, and vice versa.

A decade after Searle published his Chinese Room experiment, the Hungarian-born cognitive scientist Stevan Harnad described Searle's predicament as the "symbol grounding problem." That is, he had to answer the question "How do those Chinese characters— or, more generally, the symbols in our heads—get their meaning?" Harnad accepted Searle's refutation of strong AI on the grounds that merely receiving Chinese symbols and then manipulating them purely on the basis of their shapes did not constitute understanding, whether the input, manipulation, and output were performed by a computer or by a human non-Chinese speaker. Suppose the output of the computer and of the human being are indistinguishable from each other. We may then infer from the human being's admission that he does not *understand* a word of Chinese that the computer does not *understand* Chinese, either. As Harnad remarked in "The Symbol Grounding Problem," the meanings of a symbol system are not intrinsic to their shapes but, like the letters and words (symbols) in a book, "derive from the meanings in our heads." Their meanings are extrinsic "rather than intrinsic like the meanings in our heads"; therefore, they "are not a viable model for the meanings in our heads." In short, cognition is not simply symbol manipulation. As he points out, those who translate ancient languages or decrypt secret codes do this successfully because they are grounded in a first language and in knowledge built up from experience in the real world. This grounding is extrinsic to the symbol systems the paleographers or cryptologists are trying to translate or decode.[20]

## William James gives birth to modern "embodied cognition"

The cognitive revolution was born in a challenge to behaviorism's "black box" attitude. Before Skinner, the behaviorists asserted that the processes of cognition were "private events" and therefore inaccessible to observation. Skinner challenged this strict private event/ public event opposition but still focused on inputs and outputs rather than on cognitive processes. Some psychologists saw in the birth of computer science and AI, especially the Turing test as presented in Turing's "Computing Machinery and Intelligence," a model by which to infer a theory of cognition. If, based on input and output, it was impossible to distinguish computer "thought" from human thought, then why not assume that the processes of human cognition closely resemble the processes of computing? In short, why not conclude that both computers and human beings think?

The answer to the "Why not?" came in 1980 when John Searle persuasively demonstrated with his Chinese Room thought experiment that while both computers and human beings manipulate symbols in ways that may produce identical output, the manipulation neither requires nor implies understanding. Without understanding, there is no thought; therefore, computers, which function without understanding, do not think.

A computer is not a mind, a mind is not a computer; computation is not cognition, and cognition is not computation. But Searle left largely unaddressed the mechanism by which symbols get their meaning. Answering this is an important challenge because meaning is essential to *understanding*.

This brings us to one of the ironies of modern psychology. The mind-body dualism Descartes articulated in the seventeenth century remained influential for a very long time. Beginning in the nineteenth century, Jean-Martin Charcot, Josef Breuer, Sigmund Freud, and others created theories concerning a variety of mental processes based on observation of physical symptoms and other methods. This introduced what we call depth psychology. In the twentieth century, behaviorism challenged the inferences of depth psychology, asserting that "private events" could not be directly observed and that, therefore, mental processes were knowable only by studying input (stimulus) and output (response). By essentially equating the processes of machine computation and human cognition, the Turing test reinforced the dualism of mind and body.

Yet in his classic 1884 paper "What Is an Emotion?" the U.S. philosopher and psychologist William James published what amounts to a strong challenge to Cartesian dualism and its intellectual progeny, bringing to new birth an idea of cognition that, in fact, had been dominant for most of history. He challenged our commonsense notion that an external event creates a feeling (a mental state), which induces a bodily response: "We lose our fortune, are sorry and weep; we meet a bear, are frightened and run; we are insulted by a rival, are angry and strike." Instead, he proposed that "we feel sorry because we cry, angry because we strike, afraid because we tremble, and not that we cry, strike, or tremble, because we are sorry, angry, or fearful, as the case may be." According to James, "bodily manifestations"—weeping, running, striking—are direct responses to external events. The emotion—sorrow, fear, anger—is induced by the bodily manifestation rather than the cause of that manifestation. The body, not the mind somehow separated from the body, moves emotion.[21]

Even as Cartesian dualism continued to hold sway through the behaviorism of the twentieth century, James had already articulated what would come to be called embodied cognition, the theory that many features of cognition are influenced or shaped by aspects of the entire organism—that is, by the body. In 1984, the Polish-born U.S. social psychologist Robert Zajonc and Hazel Rose Markus, a social psychologist who had pioneered the field of cultural psychology, pursued the implications of William James's inversion of the traditionally accepted notion that emotion produces a bodily response.[22] Harnad similarly suggested that those symbols are merely projected directly into our senses: if we hold a cup of tea, the warmth that we feel from the tea is "represented" in the thermoreceptors responsible for detecting the heat. In effect, Zajonc, Markus, and Harnad all applied James's version of embodied cognition to address the symbol grounding problem, positing that we represent our emotions and perceptions by their physiological correlates. Thus, when we are happy, we do not *think* that we are happy but, rather, our zygomatic muscle becomes active—we smile—and *this*, not some thought, *is* the representation of the emotion.[23] Presumably, James would say not that we smile because we think we are happy, but that we think we are happy because we smile.

## Conceptual metaphor theory

The smile is an instance of embodied cognition, a representation of a mental state and cognitive concept we call happiness. There remains, however, an even more formidable challenge to the nature and limits of embodied cognition. How do we represent abstract concepts like democracy, justice, love, and time, for which we do not have direct physiological correlates? How do such abstract ideas acquire meaning?

We turn our focus to the highly influential conceptual metaphor theory (CMT), which was formulated by the U.S. cognitive linguist and philosopher George Lakoff and the U.S. philosopher Mark Johnson in their 1980 book, *Metaphors We Live By.*[24] CMT holds that abstract concepts are represented in concrete experiences because we coexperience them. Take the abstract concept of time, for example. We often experience time together with space. Traveling through space takes time. The coexperience of time and space is expressed in numerous metaphors that relate time and space. We may, for instance, say that we move a meeting "forward" or that the end of a boring meeting is "very far away."

For those of us who study social thermoregulation, CMT is of greatest interest when it is applied to another concrete-experience/abstract-concept pair. Lakoff and Johnson consider the relationship between (physical) warmth and affection or love. They suggest that we learn affection through experiences of physical warmth, acquiring this understanding of the symbolic concept of both "affection" and "warmth" (in its metaphorical sense of affection) by coexperiencing both affection and physical warmth when we are held affectionately by our caregivers. As Lakoff and Johnson present it, this is a clear-cut example of embodied cognition—a symbolic concept that comes to be understood through bodily experience.

Lakoff and Johnson propose that the cognitive metaphor here flows in one direction only. That is, one who experiences affection often coexperiences physical warmth, so that physical warmth becomes available as a metaphor of affection and creates a bodily, grounded understanding of affection. However, just because you happen to be warm (standing out in the sun, say) or cold (standing out in the gloomy snow) does not mean you will necessarily *experience* affection or its absence. By this one-way logic, if you turn up the thermostat in your room, you should feel more affectionate, but if you happen to activate affection, you will not necessarily experience more warmth.

I have a hard time uniting Lakoff and Johnson's approach with what I know about social thermoregulation. As we saw in Chapter 1, at least some of the studies with hot and cold beverages and foods do not support this one-way view. Recall, for example, the University of Toronto experiment in which people who were ignored in the Cyberball video game—did not get the ball passed to them—and were thus made to feel rejected favored a list of hot foods and beverages more than did those who felt included because the ball had been tossed to them frequently. There were no preference differences between rejected and included volunteers when it came to temperature-neutral control foods and beverages such as Coke or apples. The opposite of affection—that is, exclusion—induced a preference for physical warmth. This result is not good news for CMT. If a conceptual metaphor were at work here, we would expect that exclusion or feeling lonely would not have affected our feelings of temperature. But it did, and it also drives our preference for becoming warmer.

You may also recall from Chapter 1 that two of my own studies were consistent with what the Toronto researchers found. If people were made to feel different from others, they judged air temperature to be lower. If they were made to feel alike, they judged the temperature as higher. If they *read* about "warm" people—people who are loyal, friendly, helpful—they also judged ambient temperature as warmer. This is not some kind of metaphor from Descartes's "pilot" in the skull but a much more dynamic social thermoregulation.

Think also about the study my colleagues and I did in Sopot, Poland. Volunteers who had read about characters associated with "warm" qualities (recall that we did not use this word) experienced the air temperature as more than 3 Fahrenheit degrees higher than those who read about characters described with terms that, although positive, did not imply emotional or social "warmth." If a metaphoric process were operating here, feeling physically warm might evoke emotionally or socially "warm" thoughts, but Lakoff and Johnson would have predicted that these thoughts would not work in the other direction, to make a room seem warmer. In their view, the mere feelings of affection should not affect thermoregulation.

Lakoff and Johnson were correct about the nature of metaphors. They work in one direction only. When the Scottish poet Robert Burns sang out, "O my Love is like a red, red rose," he used the rose as a metaphor that evokes the passion he feels for his "bonnie lass."[25] But try to reverse the metaphor, and you get nonsense. "O a red, red rose is like my Love" suggests that Burns has a passion for roses equal to or even greater than what he feels for his girlfriend. Unless Burns was a very strange fellow indeed, the metaphor must function in one direction only. The point of this line of poetry is to describe the poet's passion for his love, not for a rose. The Cyberball and Toronto studies, along with my own studies discussed above, reveal that the association between physical temperature and emotional/social "temperature" works in both directions. This bidirectionality implies that the basis of the association cannot be metaphorical.

Their reliance on metaphor requires Lakoff and Johnson to establish another central tenet of CMT, namely that the concepts described are not innate. This contradicts the fact that the need for social thermoregulation *is* innate. If the concepts described are not innate, the metaphor—which Lakoff and Johnson call a "primary metaphor"—*must* adequately serve the purposes of an innate property. That is, the metaphor in CMT *must* be universal. Yet we know that it is certainly not universal, because the language of the metaphor varies widely, as we will discuss in Chapter 5. If CMT has proved inadequate to account for social thermoregulation because it ultimately forces upon us a dualism of mind and body, we need an intellectual revolution that begins with a better theory of how people engage in social relationships. We will start moving in this direction, but the route will take us to human society via a detour to a bird named Harry the Penguin.

## CHAPTER 3

# Harry the Penguin

## *How Animals Deal with Temperature*

Imagine an emperor penguin, about 10 years old, male. Let's name him Harry. It's July, the middle of the Antarctic winter, and Harry is into his fourth month on an empty stomach. Male emperor penguins routinely fast when they pair with females, and they continue without food for as long as 115 days. After mating, the females lay eggs and then waddle back to sea in search of food. They won't be back for months, having left each stay-at-home penguin dad with an egg to care for.

So, the wait begins. Harry places his egg on his feet and envelops it in his brood pouch, a thick, skirtlike layer of naked skin. To ensure incubation, the egg is kept at a cozy 96.8°F (36°C), not easy in Antarctica, where winds can blow at 110 miles per hour and temperatures may dip below −49°F (−45°C). If Harry wants his egg to survive—and if he wants to go on living himself—he must huddle with others.

There are plenty of photos and videos from Antarctica featuring emperor penguins standing as close together as rock fans at a sold-out concert. Thousands of individuals may pack tightly, leaving each bird no more than a square foot to himself. The pressure inside the group can become so great that some penguins are pushed up on top of the others. If this really were a rock concert, they'd almost be crowd surfing. Yet as a means of saving energy, all this apparent discomfort makes perfect sense. When penguins huddle, they drastically limit the amount of body surface each of them exposes to the elements, making it far easier to conserve heat. What is more, each of the penguins warms up the microclimate of the group. As a result, the temperature inside the pack may rise to 99.5°F (37.5°C).[1]

For years, in places like Pointe Géologie, the rocky archipelago off the coast of Antarctica, scientists have been conducting studies of huddling penguins. To better understand why and how the animals huddle, researchers glue temperature-measuring instruments and transmitters to the feathers of selected birds. Such studies have shown, for example, that the penguins may spend as much as 38 percent of their time packed tightly with others, sometimes huddling for just a few minutes at a time and sometimes for hours on end.[2] The wind-whipped group may appear perfectly still, but time-lapse videos show that the structure of the group constantly fluctuates as penguins move within it, swapping places with one another.

At first glance, the explanation for why emperor penguins huddle seems intuitive. If the proverbial chicken crosses the road "to get to the other side," penguins huddle "to keep each other warm." Seems like a no-brainer. But experiments that involved putting penguins on treadmills have shown that huddling is about far more than just staying snug. Huddling is about hunger, survival, and, on a social level, the coherence of the group.

After one study conducted at Ross Island, Antarctica, researchers concluded that if penguins did not huddle, they would starve. The math is rather simple. To endure a fast of 100 days, a penguin would have to use up about 55 pounds of fat tissue. He would also need an additional 3 pounds of fat for the long walk—as much as 125 miles—back to the sea, where all the nutritious fish and squid are to be found. Meanwhile, a large emperor penguin such as our Harry may have only between 33 and 44 pounds of fat tissue in reserve. In other words, to survive the long fast and the long walk, he is short at least 14 pounds of fat. The good news? Huddling, and the warmth it conserves, allows Harry to turn down his metabolism by as much as 16 percent and thereby conserve much-needed fat.[3]

## The economy of action

Penguins are not the only animals that use others to thermoregulate. Rats, pigs, monkeys, horses, snakes, woodpeckers, raccoons, and even porcupines—not the first creature you would think of cozying up to—huddle when the mercury goes down. In fact, many animal species do. Next only to getting enough oxygen, temperature regulation is the most vital and energetically expensive thing an animal must do. Unlike oxygen, temperature fluctuates in our environment all the time, and our survival therefore requires continual vigilance to deal with the fluctuations. Fortunately, it turns out that animals are very canny economists or, perhaps better, "optimizing agents." They continually evaluate the cost versus the benefit of different behaviors, in an effort, it seems, to figure out which ones will prove the cheapest from the perspective of energy and thereby help conserve precious body fat.

If you've ever seen a group of cyclists racing on the road, you might have noticed that they tend to follow others in an elongated pack, called a peloton. It's not that the cyclists are lonely. They are relying on one another, "drafting" to conserve energy by exploiting the lower air-resistance behind other racers. Skilled drafters can reduce their own power requirements by as much as 39 percent at a speed of 24 miles per hour, especially if they are in the middle of the peloton.[4] This is also why fish swim in schools, and even why bacteria travel faster as a group. Like huddling, drafting is all about something called the "economy of action": choosing the most cost-efficient behaviors to save resources. The economy of action is about a stunningly simple principle: animals need to take in more energy than they spend. If they don't, they die.

Across species, orders, and phyla, animal economists have developed an astounding array of mechanisms to deal with temperature regulation and energy savings. Which specific tools an animal has at its disposal depends largely on what type of organism it is. If you went to school a few decades ago, you might have been taught that we divide animals into warm-blooded and cold-blooded, and you might be surprised to learn that this distinction, once so commonplace, is now considered obsolete. Today, scientists prefer to use other terms, admittedly more challenging to the tongue: *ectotherms*, *endotherms*, *poikilotherms*, *homeotherms*, and *heterotherms*.[5] The argument for this plethora of new vocabulary is that the cold-blooded-versus-warm-blooded division was not only a vast oversimplification but also often quite confusing.

Consider a lizard basking in the August sun somewhere in Texas or Arizona. Traditionally, we would have said that the lizard, as a reptile, is cold-blooded. But while basking, its core body temperature can rise to a sweltering 100.4°F (38°C), which is quite far from having "cold" blood. Enter *ectotherm*, the new go-to classification for our lizard. Ectotherms are defined as animals that derive their body temperature from their environment but may also optimize metabolism and their performance by engaging in deliberate thermoregulatory behaviors. In general, those animals that were formerly labeled "cold-blooded" are today's ectotherms. Think insects, reptiles, amphibians, and most fish. In contrast, mammals and birds, animals we used to call "warm-blooded," can generate their own heat, as if they have an internal furnace. Today, scientists call them endotherms. *Ecto-* and *endo-*, both from Greek roots, mean "external" and "internal," respectively. *Therm* is from the Greek word for heat, *thermē*.

Relatedly, to be a *poikilotherm* means to have a body temperature that changes with the environment (from the Greek *poikilos*, "changeable"), and to be a *homeotherm* means to keep a high and stable body temperature no matter the weather (the Greek *homoios* means "similar"). *Heterotherms*, meanwhile, are somewhere in between poikilotherms and homeotherms. They sometimes have stable body temperatures but sometimes let go and allow their body temperature to yo-yo freely. A giant Nile crocodile weighing over 1,100 pounds is essentially an ectotherm (no internal central heating), but its enormous mass also makes it a homeotherm because its body temperature remains fairly stable no matter how hot or how chilly it is outside. Large mass makes for large thermal inertia. It is the same reason why it takes forever to cool down or warm up water in a big pot. On the other hand, the body temperature of some endotherms can vary greatly over the course of a single day, sometimes by as much as 104 Fahrenheit degrees (40 Celsius degrees), as is the case with one species of hibernating arctic squirrels. This makes such mammals poikilotherms.

If you want to be truly accurate, you might say that humans are endothermic homeotherms who may sometimes be a bit poikilothermic, heterothermic, and even ectothermic. Accurate? Maybe. A mouthful? Absolutely. Thus, to keep things more manageable, I'll mostly use the terms *endotherms* and *ectotherms* and go for the other ones only where absolutely necessary.

## Slacker or investment banker?

At first glance, being an ectotherm looks like a bad deal compared to being an endotherm. Incapable of independently warming their own bodies, ectotherms live a weather-dependent life, whereas endotherms have more autonomous control and flexibility. When the mercury falls, the body temperature of a snake or a frog falls accordingly. If a tiny fly lands in a spot of sunshine, it may heat up by 50 Fahrenheit degrees (10 Celsius degrees) in a mere 10 seconds. No wonder that, a century or so ago, scientists considered ectotherms inferior to endotherms, less evolved than "warm-blooded" monkeys, dogs, and humans. But more recent research shows that neither approach to thermoregulation is inherently better than the other. Each has advantages and disadvantages.

You might think of ectothermy and endothermy as two contrast-ing lifestyles. Picture a slacker leading the slow life in the Florida Keys, selling the occasional T-shirt to tourists, and then picture a fast-tracking investment banker on the hustle in New York. In this scenario, insects and reptiles are like the slacker. Their ectother-mic lifestyle is the slower, lower-energy approach to life. Mammals and birds are the investment bankers, whose endothermic lifestyle is a fast, high-energy existence. If you want to tightly control your body temperature, you need to fuel your internal furnace—your metabolism—and that means eating a lot. Getting a lot to eat means that you must spend a good deal of time chasing your next meal instead of just lazing about. Letting their body temperature fluc-tuate with the environment allows ectotherms such as reptiles and amphibians to survive on much less food than a mammal of equal size. A rodent weighing 10.5 ounces would have to eat a staggering 17 times more insects per day than a lizard of exactly the same size. Ectothermy is also the reason why some snakes, such as pythons, can go a whole year with no food at all. And don't assume that endo-therms are independent of climate and weather. The colder it gets, the more food mammals need to keep their internal furnace going. Female mice, for example, cram in roughly 38 percent more food at 71.6°F (22°C) than they do at 86°F (30°C).

As slackers living the slow life, ectotherms tend to have offspring later in life than endotherms do. That gives them more time to grow before they reach maturity. In addition, because they can boost their metabolism by using heat energy from their surroundings—instead of relying exclusively on food—ectotherms may be even *more* flex-ible than endotherms when it comes to finding compatible environ-ments. To achieve similar flexibility, those fast-living investment bankers, the endotherms, need to "invest" in their group. The pru-dent endotherm always broadens his "investment portfolio." Instead of relying exclusively on the food he can find on his own, he invests some of his independence in social thermoregulation. He relies on the energy derived from the warm bodies of others. In fact, the more he can consistently rely on those other bodies, the better. Harry the Penguin is an investment banker, investing heavily in his rookery mates, huddling with them and living off the dividends he receives from their radiated heat even as they profit from his. In the social compact that is huddling, Harry and his rookery mates "spend" some independence to purchase the mutual benefit of heat energy.

Despite the high energy demands of endothermy, which restrict environmental flexibility and individual independence, the fast-track lifestyle does have an impressive upside. At high temperatures, chemical reactions proceed more rapidly, and relatively stable body temperatures allow enzymes to function more efficiently. As a result, mammals and birds can sustain relatively higher levels of activity than amphibians or reptiles. The difference is most pronounced when ambient temperatures are low. On some especially chilly day, go visit a friend who owns a tortoise. You will see just how sluggish it can be. Yet, at the height of summer, that same reptile may be able to run—yes, *run*—at surprisingly brisk speeds. Warm muscles contract faster, which also explains why houseflies are at their most annoying on those toasty, picnic-perfect days.

## When endotherms and ectotherms share thermal strategies

Even though endotherms can regulate their body temperatures on the inside—more on the specific mechanisms in the next chapter—their preferred method of thermoregulation is, in fact, the same one that ectotherms favor: to change behavior. They bask in the sun, hide in burrows, hunch, and spread their wings. Some even hug trees. Others, like Harry the Penguin, huddle.

Every morning around seven, a small male lizard of the *Liolaemus multiformis* species emerges from its burrow, where it has been hiding from the midnight chill. Of all lizards, *Liolaemus multiformis* lives at what is by far the highest altitude. It inhabits the Andean grasslands of Peru, dwelling at dizzying heights of over 11,400 feet. In this challenging environment, nighttime temperatures can fall below freezing. To start the day, the lizard must warm up, overcoming its cold-induced sluggishness. Looking around, it spots a sun-kissed mat of vegetation, a place well insulated from the icy ground. It lies there, basking in the sunshine. Within just two hours of waking, by nine, the animal's body temperature has risen to 95°F (35°C), which allows it to hunt and digest quickly. *Liolaemus multiformis* is ready to go! If conditions remain cloudless, the lizard will be able to maintain its high body temperature throughout the day, with little or no energy cost.

Plenty of animal species use sunshine to regulate their body temperature on the cheap. Ectotherms such as lizards, frogs, and butterflies, as well as endotherms such as seals, penguins, and lemurs, all make use of the sun. In their natural habitat of the southern Rocky Mountains, newly metamorphosed western toads climb on top of plants to profit from solar rays. When locked away in research labs and lacking sunshine, the tiny amphibians tend to sit beneath incandescent lamps. Either way, they warm themselves to their preferred body temperature of about 80°F (26.7C), which in turn allows them to grow faster. In an interesting twist, if their stomachs are empty, the little toads go for a lower body temperature, just 59°F to 68°F (15°C to 20°C), a state far easier—and therefore cheaper—to maintain. These toads are governed by a basic form of the economy of action. If food is abundant, the toads allow themselves a higher body temperature, enabling their bodily enzymes to work more efficiently. If food is scarce, they go into energy-saving mode.[6]

Your parents may have scolded you for slouching, but slouching—as well as other postural changes—is another cheap way some animals regulate body temperature. Monkeys, lemurs, rodents, and seals have all been observed hunching to reduce their surface-area-to-volume ratio, thereby slowing down heat loss when it's cold outside. If excessive heat is the problem, the strategies are different.

The springbok, a medium-sized African antelope, can reduce the amount of solar radiation reaching its body by a whopping 62 percent, according to some researchers, simply by standing parallel to the direction of the sun or lying down in the same orientation, shading itself with its own bulk. So, if you ever get stranded in the desert, consider adopting a springbok pose to enhance your chances of survival. Of course, an even better idea is to hide somewhere out of the sun. Small desert animals, for which regulating body temperature is especially costly and difficult, typically spend scorching days hidden in burrows, emerging only at night.

For those animals that don't much like burrows yet are still in need of ready heat-relief, tree hugging is an option. Researchers in Australia have observed a surprising behavior in koalas. In hot weather, they spend a considerable amount of time on *Acacia mernsii*, trees on which they do not feed. The koala hugs the trunk, its furred belly pressed snugly against the bark. When scientists calculated the temperature of *Acacia* trees, the rationale for this behavior became apparent. The tree trunks are typically cooler than the surrounding air by an average of 41 Fahrenheit degrees (about 23 Celsius degrees),[7] enticing the koalas to use them as a cold compress.

In case cool trees and shaded burrows are not sufficient, an overheated animal can go for something called evaporative cooling: sweating, panting, and licking. It takes about 500 kilocalories to evaporate a single liter (a bit more than a quarter of a gallon) of water, whether the water is from sweat or saliva. There is no more effective way of ridding oneself of excess body heat. Cats lick themselves to cool down and so do mice. Vultures, for the same reason, pee on their legs. As the urine evaporates, it draws out the heat from the bird's body. Dogs don't lick themselves as much as cats do, and if they pee down their leg, it's likely not for temperature regulation. They are masters of another evaporative cooling technique: panting. The frequency of their breaths increases over 10 times, enabling the loss of moisture through the nose, trachea, bronchi, mouth, and tongue.

If you try to pant, you will soon discover that it is tremendously difficult. That's because we humans cannot truly pant. Instead, we hyperventilate and pass out. Panting animals—endotherms such as dogs, sheep, cows, birds, and even ectothermic lizards—increase ventilation mostly in the upper respiratory tract, without much additional air exchange happening in the lungs. What is more, they make their respiratory system oscillate at its resonant frequency, which reduces the strain on the muscles.

In general, smaller animals are masters of panting, and larger ones masters of sweating. One possible exception to this generality is very large dogs, such as Great Danes and Irish wolfhounds. Although much bigger than Chihuahuas, they still pant. Dogs, by the way, do sweat, though minimally, and almost exclusively through their noses and the pads of their paws. Birds do not sweat at all, which is a good thing, considering that wet feathers would make for very difficult flying.

Endotherms are not the only animals that can sweat or pant. On a particularly hot night, as you lie perspiring miserably in your bed while cicadas sing merrily outside, be aware that those little ectotherms may be sweating, too. On such a night, even frogs may sweat. Although cicadas and frogs don't have anything like mammalian sweat glands, some species do secrete watery mucus from their skin.

Of course, what helps even more to keep a frog cool is being buck naked. Even though some of us find the prospect of a woolly frog cute, there is a good reason why such creatures have not evolved on Earth. As ectotherms devoid of an internal temperature-regulating system, an internal "furnace," frogs, lizards, and snakes must be able to efficiently transfer heat between their bodies and the environment. Fur and feathers slow down such transfer. For endotherms, that's a good thing. Just as a house equipped with central heating is better off with good insulation to lower fuel costs, a mammal or a bird lowers its heating costs with a well-insulated body.

Fur and feathers are not the only effective endotherm insulation material. Fat is another. It does not transfer heat as readily as muscle or skin. A camel's humps work so well because of fat's insulating properties. If these animals had nothing more than a layer of fat spread under the skin across the body, as we humans do, they would not be able to cool down efficiently. Humps are a compromise. The fat stores are there in case of food shortages—not uncommon in the desert. Camel fur is an economic compromise of sorts, too. Although common sense might suggest that naked skin would be of benefit to desert-dwelling animals, some types of fur offer needed protection against radiant heat. That is also probably why we humans, mostly hairless creatures, still have hair on top of our heads. Like a hat, it shields the very part of our body most exposed to the sun.

Once the sun goes down, we might fluff up our hair to better protect us from the chill creeping in. Just as birds plump up their feathers when it's cold outside, mammals other than humans can profit from piloerection, in which special muscles at the base of hair or feathers activate and raise them up. This way, the amount of air trapped in the pelage (the zoological term for the hair, fur, or wool of a mammal) or feather cover increases, and so do its insulating properties. Since many of us have little hair on our bodies, human piloerection manifests itself primarily as goosebumps on naked skin.

Although fluffing fur or feathers, panting, and hugging trees are all great methods for regulating body temperature, endotherms possess an additional set of tools—unavailable to ectotherms—to keep themselves in thermal comfort. They can generate their own body heat. One way to do this is to shiver, as you might do when you step out of the cold sea and onto a windy beach. When you shiver, some especially fatigue-resistant muscles contract rapidly and involuntarily, thereby generating heat. In some animals, this can go on for weeks at a time. All mammals and birds shiver when they feel the chill, penguins and arctic foxes included. Don't bother, however, to look at birds on a wintry day to see them shiver—bird anatomy makes the vibrations invisible.

Besides shivering, endotherms can warm themselves by firing up their metabolism. As Harry the Penguin, battered by Antarctic winds, waits for his egg to hatch, he burns through his fat stores to keep his body temperature up and stable. By the end of the season, he will have lost a large chunk, roughly 40 percent, of his initial weight. (What birds, Harry included, do not have is something called brown adipose tissue [BAT], a specialized tissue found in some mammals, such as humans and rodents, that is especially good when it comes to generating large amounts of heat. We'll have more to say about BAT in Chapter 4.)

## Hibernation, torpor, and huddling

Producing your own heat is very costly, and shivering has been found to be a pretty inefficient method of doing this. Humans usually turn to shivering only as a last resort. If you start to shiver, your core body temperature is likely already affected. To conserve energy, some endotherms draw on a few special strategies instead. Foremost among these are hibernation and torpor. Back in 1987, 12 Arctic ground squirrels were placed in outdoor wire cages in Fairbanks, Alaska. All the animals had miniature temperature-sensor radio transmitters implanted in their bodies, which allowed scientists to remotely monitor their vitals in real time. The animals dug burrows, and awaited winter. When it came, it was long and glacial, typical of Alaska, with temperatures dropping below −30°F (−34°C). The squirrels went into hibernation. That is when something truly unusual began to happen. Their core body temperature kept falling and falling—below freezing, reaching a minimum of 26°F (−3.33°C). Yet, come spring, they had survived to awaken to another season. What's more, they likely saved 10 times more energy compared to what they would have used up if they had maintained their body temperatures at just above freezing.[8]

Although Arctic ground squirrels probably offer the most dramatic example of hibernation, made possible by unique antifreeze molecules in their blood, many other mammals survive the cold by essentially putting life on hold, dramatically decreasing their body temperature and becoming immobile, all to save energy. If an animal does this for short periods of time, say, from nightfall till dawn, it's called torpor. Hibernation is a prolonged form of torpor, with body temperature generally falling to just slightly above air temperature. Mice hibernate, and so do bats, hamsters, and hedgehogs; while skunks and lemurs, and even some birds such as hummingbirds and insect-eating swifts, enter torpor.

The "hibernation" of bears is familiar to everyone—except that bears are not true hibernators. No large mammals are. The bear's body temperature falls just a little bit, and its metabolism and bodily functions don't slow down much, either. That's why scientists avoid the term *hibernation* when speaking of bears and say instead that they enter a "winter sleep." If bears were true hibernators, strolling into their caves in the middle of winter would be quite safe. As things are, it's best avoided. Really.

Hibernation and torpor are great ways to save energy. Animals enter hibernation voluntarily, primarily to save energy and especially when food is scarce, as in the winter. Hibernation periods may last from several days to several weeks, during which animals do not forage for food. Therefore, before entering hibernation, they stock up and then try to stay in the low-energy hibernating state as long as possible. In contrast, animals enter into torpor involuntarily, often for short periods of less than 24 hours. Depending on the species and external conditions, the animal may even enter torpor daily. Daily torpor is accompanied by continued foraging. Torpor is usually triggered by temperature fluctuations and reduced availability of food. Some scientists believe that hibernation and torpor are not radically different states but merely extremes along a continuum. It may be helpful to think of torpor as "hibernation lite"—especially if doing so discourages you from entering the cave of a bear in torpor.

Keeping core body temperature dramatically or even relatively low slows metabolism, which in turn means burning more slowly through vital fat reserves. When a mouse has trouble finding food—

say you've closed up your pantry tight—its body temperature plum-mets, and it will hibernate, waiting out the tough times. The animal's metabolic rate may go down by as much as 75 percent, perhaps even more. Precious energy will be saved. To put that savings into per-spective, a hibernating animal can survive 40 times longer on its fat reserves than an active one.

Torpor and hibernation are not necessarily the ultimate energy-saving strategies. Some animals save even more energy by hibernating while huddling with others. As social animals go, marmots are cer-tainly at the very top of the list. Himalayan marmots, fluffy animals that could model for a cute stuffed toy, tend to play-fight together, communicate by loud whistling, and greet each other by touching nose to nose. When it's time to hibernate, they do so as a group.

Alpine marmots hibernate in sync, and they awaken in sync, too. Those marmots that are temporarily awake while the others are still in hibernation tend to stay huddled with the rest and may even groom them or cover them with hay. Thanks to their friends, yellow-bellied marmots, a species that inhabits North America, can save as much as 44 percent of their energy.[9] The profits derived from hibernating together are so great, researchers now believe, marmots have become socially synchronized to enable them to huddle while hibernating. In evolutionary terms, the reasoning goes that synchro-nization was initially all about the care of offspring. Since marmots live in harsh environments, they need to get bulky and fat to be able to survive winters, despite the short growing season. The young ones generally don't manage to grow enough over their first summer and therefore remain highly vulnerable when the snows come. Hibernat-ing with their relatives allowed them to survive even with low stores of body fat. From there, the habit of huddling together in winter spread to include adult animals from outside the immediate family. Everyone profited, and the huddling behavior caught on—in an evo-lutionary sense.

Social thermoregulation may also be the reason why rodents tend to live in groups. If you have seen Siberian hamsters in a pet store,

you likely saw them cutely piled up. Rodents are very much into huddling. Many species, from house mice and red-backed voles to flying
squirrels, engage in this behavior. There are now numerous studies
suggesting support for the theory that group living evolved in rodents
precisely because it allowed the animals to save energy required to

### Metabolic (energy) savings from huddling (in %)[10]

| | |
|---|---|
| African four-striped grass mouse *Rhabdomys pumilio* | 16 |
| Bank vole *Clethrionomys glareolus* | 8–35 |
| Emperor penguin *Aptenodytes forsteri* | 16 |
| Antelope ground squirrel *Ammospermophilus leucurus* | 40 |
| Grey partridge *Perdix perdix* | 6–24 |
| Muskrat *Ondrata zibethicus* | 11–14 |
| Harvest mouse *Reithrodontomys megalotis* | 28 |
| Green woodhoopoe *Phoeniculus purpureus* | 30 |
| White-backed mousebird *Colius colius* | 50 |
| Townsend's vole *Microtus townsendii* | 16 |
| European common vole *Microtus arvalis* | 36 |
| Striped field mouse *Apodemus agrarius* | 12–29 |
| Tundra vole *Microtus oeconomus* | 10–15 |
| Common bushtit *Psaltriparus minimus* | 21 |
| Gray mouse lemur *Microcebus murinus* | 20–40 |
| Australian hopping mouse *Notomys alexis* | 18 |
| Naked mole-rat *Heterocephalus glaber* | 22 |
| Domestic rabbit *Oryctolagus cuniculus* | 32–40 |
| Common rat *Rattus norvegicus* | 34 |
| House mouse *Mus musculus* | 14–22 |
| Yellow-necked field mouse *Apodemus flavicollis* | 13–44 |
| Northern white-footed mouse *Peromyscus leucopus noveboracensis* | 27–53 |
| Lesser bulldog bat *Noctilio albiventris* | 38–47 |
| Red-billed woodhoopoe *Phoeniculus purpureus* | 12–29 |
| Speckled mousebird *Colius striatus* | 11–31 |

regulate their body temperature.[11] Huddling, after all, is more common in cold habitats (think *Siberian* hamsters), and burrow-sharing increases in winter.

Plenty of studies have shown that huddling allows animals to ratchet down their metabolism. Researchers have calculated, for example, that for a Chilean rodent called the common degus, huddling allows for about a 40 percent reduction in basal metabolic rate. What's more, the effects persist even after the winter, when the animals no longer huddle. Such reduction in basal metabolism also allows the degus to eat less. The same thing happens with mice. If you hold a mouse alone in a cage kept at 71°F (21.67°C), it will eat a staggering 22 percent more than if it had two other cage mates. (Interestingly, human beings in committed relationships consume less glucose, but whether this has anything to do with hugging—an analogue of huddling—remains to be answered.) As shown in the table on the previous page, social thermoregulation allows animals to save between 6 and 53 percent of their energy, depending on the species and the size of their group.

## Society and survival

Social relationships and the lower energy needs that often go with them can make survival easier in times of scarcity. Take what happened to Tony and Greg, two wild Barbary macaques. The winter of 2008/2009 was exceptionally harsh in the Middle Atlas Mountains of Morocco, where Tony and Greg lived. The ground here, at the very edge of the Sahara Desert, was covered with as much as 35 inches of snow, and the temperatures plummeted. Roads were closed for weeks. During that particular winter, a group of researchers observed the fate of 47 wild Barbary macaques, including Tony and Greg. They calculated how much time each monkey spent foraging and how many social relationships he or she had, based on exchange of grooming or simply touching each other. By the end of winter,

30 of the macaques had died of hunger, unable to find enough food under the deep snow. The researchers noted an unusual pattern, however. Those animals with more social connections had better odds of making it through the winter. Each additional relationship yielded a 0.48 increase in the odds of survival. Alas, Tony and Greg, who were exceptionally asocial, died.[12]

Although this study did not directly calculate the impact of huddling on energy needs and survival, other studies confirm that monkeys with abundant social partners have body temperatures that do not fall as low as those of solitary animals. What is more, they prefer to huddle at night with those friends with whom they interact the most during the day. If such interactions involve grooming, all the better.

Grooming in primates has been interpreted as a means of removing parasites, and as a social behavior. Monkeys groom each other even in the absence of infestation and do so at rates that are higher than required by hygiene alone. New experimental evidence suggests that grooming can be not only social but also, more specifically, about social thermoregulation and energy conservation. Vervet monkeys, a species native to southeastern Africa, are often studied for the similarities they exhibit to human social behavior. Richard McFarland at the University of Wisconsin–Madison and his colleagues obtained the pelts of seven adult vervet monkeys in the Eastern Cape of South Africa. (All had died from natural causes.) After the pelts were tanned by a taxidermist, the researchers back-combed them as a proxy for grooming. They discovered that the back-combed pelts became deeper and trapped more air, giving them better insulating properties. A well-groomed fur would also likely be free of matting, allowing for higher efficiency in piloerection, when fur is raised to make the coat warmer.[13]

Primates with stronger social relationships—the better-groomed animals with more huddling partners—incur lower energetic costs in regulating their body temperature. They have lower nutritional needs and higher chances of survival. British anthropologist Robin

Dunbar, who has extensively studied the role of grooming in primate societies, is well known for his theory that grooming serves to build relationships.[14] But it is even more basic than even Dunbar had assumed. It appears that a groomed monkey "knows" it can rely on others to keep warm. This, in turn, encourages the animals to live in larger and larger groups. Sociality flourishes.

## Endo? Ecto? Sometimes it depends on the company you keep

Naked mole-rats are, frankly, ugly creatures. They have pink, hairless skin so wrinkled that it looks too big for them. They have large, protruding teeth, tinted an unappealing yellow. Their bodies resemble fat sausages.

But it's not just their looks that make them, well, *special*. The naked mole-rat is the only mammal that is poikilothermic. If the air temperature declines, the body temperature of a naked mole-rat declines commensurately. If the mercury ascends, the temperature of the animal also goes up, much as it would in a lizard or a frog. The likely motive for the mole-rat having given up on mammal-style thermoregulation is its energy costs. The disturbing-looking rodents live in northeast Africa in extensive mazes of underground burrows. Finding food in the soil is far more difficult than finding it on the surface. The energy costs of coming across something edible are 4,000 times higher below ground than above. So, to reduce their energy spending, the naked mole-rats gave up their internal furnaces—a great way to lower heating costs.

But there's a plot twist. A single naked mole-rat may be a "cold-blooded" poikilotherm, yet put a few of them together, and they suddenly turn into much more typical mammals. Tightly huddled, a group of naked mole-rats is no longer poikilothermic. The group is homeothermic. The internal temperature of each member becomes quite elevated and stable, much like that of an average dog or cat.

When naked mole-rats huddle, it is as if they merge into one "warm-blooded" super rat.[15]

For an explanation of this transformation, let's look not to biology but rather to physics. Passive heat loss depends on the ratio of surface area to volume. The higher the ratio, the greater the heat loss. That is why small animals, which have a higher skin-to-mass ratio than do large mammals such as elephants or rhinos, are far more likely to suffer when the temperature goes down. Yet when animals huddle, their total surface-area-to-volume ratio goes up considerably, and each individual in the pack exposes less of itself to the elements. When rodents huddle, they reduce the body surface that is out in the air by between 29 and 39 percent. This is especially important for naked mole-rats, since their hairless skin is so poorly insulated.

A huddling superorganism creates its own microclimate, too. Every individual in the heap dissipates warmth to its surroundings, and the overall temperature within and in the immediate vicinity of the huddle rises. Admittedly, not every spot in the huddle is equally good. If you've ever been to a rock concert, you might have noticed that it is much hotter in the very center of the crowd than it is on the edges. The same thing happens when animals huddle. As if to avoid discrimination, individuals may periodically swap places.

One exception to this reciprocation of social thermoregulation can be found in marmots. Alpine marmots engage in a practice called alloparenting, which is a type of care for descendants that are not directly one's own (but can be a grandchild or a niece). Having a larger group increases the chances of survival for the parents and their young. But marmots distinguish between dominant and subordinate adults when socially thermoregulating: the subordinate marmots stay on the outside of the huddle, thereby increasing their chances for their early death.[16] Who knew that marmots could be such exclusive jerks?

In one experiment, researchers marked common rat pups on their backs and used time-lapse videography to observe how the young rodents moved around in the huddle. The researchers referred to this movement as a "pup flow." When it was cold, the general movement was downward, with "convection currents" of small, drowsy bodies passing through the clump. In a warm nest, the overall direction of the pup flow was upward. What's more, the huddle of young rats behaved, once again, like a superorganism, adjusting its collective shape depending on changes in the ambient temperature—expanding when the mercury went up, and contracting when it plunged.[17]

## Ectothermic thermoregulation

Mammals and birds are not the only animals that engage in social thermoregulation. Such ectotherms as insects use it, too. Some species of ants, such as neotropical army ants—giant ants that reach as much as a half inch in length—can huddle together for the sake of their brood. This is called an ant "bivouac" (these *are* "army" ants, after all), which has a stable, cozy temperature within its core, perfect for larvae and pupae. When it turns cold outside, the bivouac changes shape, becoming more rounded to reduce the surface-area-to-volume ratio and thereby preserving more heat.

In contrast to huddling army ants, honeybees shiver to warm up their nests, all the workers contracting their flight muscles in perfect synchrony. The results can be highly impressive. A colony of honeybees keeps the brood-nest temperature stable at a toasty 91°F to 96°F (32.77°C to 35.55°C), even when the air temperature is below freezing. Japanese honeybees derive even more unusual benefits from social thermoregulation. From time to time, their nests are attacked by *Vespa mandarina japonica*, the rather scary Japanese giant hornets famed for being the largest hornets on earth, often measuring over 1.8 inches in length. A sting from one might land you in the hospital. As a weapon against such dangerous predators,

Japanese honeybees use body heat. The workers cluster together around the nest, shivering to increase their body temperature. When it reaches 114.8°F (46°C), it becomes lethal for the hornets, but not for the honeybees.

Snakes may not use heat as a weapon, but some of them do use it in ways that we humans, at least the ladies and gentlemen among us, might find unethical. They steal it. Male garter snakes in Canada, for instance, pretend to be female by producing female-like pheromones so that other males will try to mate with them. As many as one hundred snakes can climb on top of such a deceitful male, warming his body, under false pretenses, for free. Amphibious sea snakes of New Caledonia also engage in heat theft, or what scientists call kleptothermy. They may sneak into burrows occupied by large tropical seabirds, taking advantage of the massed body heat that warms the space. In similar fashion, lizards, snakes, and even dwarf caimans freeload on the heat produced by termite mounds. Native to the Amazon, Schneider's dwarf caimans (which at 7.5 feet are not exactly dwarfish) place their eggs beside or on top of termite mounds to ensure that they stay warm.

## Social thermoregulation as an energy bargain

Maintaining optimal body temperature is an imperative task for any animal. Being too hot or too cold can have dire consequences. As when Goldilocks pilfered from the three bears, the porridge must be "just right."

Both ectotherms and endotherms have an array of tools to help them optimize body temperature—from sunning and panting to sweating and hiding in burrows. Mammals and birds can also rely on their internal heating systems to maintain their core body temperature in its most advantageous range. The problem with the endotherm tools, however, is that they are costly in terms of energy.

Some, like shivering, are costlier than others. Thus, following the principle of economy of action, animals are always bargain-hunting and on the lookout for opportunities to cut costs.

Enter social thermoregulation. Stealing warmth, grooming each other, or huddling together to stay snug, animals have evolved an array of social behaviors aimed at thermoregulating. As we've seen in this chapter, fascinating examples abound throughout the animal kingdom. If you have relatives or friends nearby, there is a good chance that they will keep you warm and make it less energetically costly for you to live. You may need less water, less food. You may grow faster. Having someone to count on as a source of heat can mean the difference between life and death. Thus, it is vital to know how socially reliable those around you are.

Like an investment banker on life's fast track, an endotherm such as Harry the Penguin forms relationships with his rookery mates to make his environment more stable and predictable. The more coherent his group, the more likely he is to survive.

Yet there is also asymmetry in social thermoregulation. In any of its forms, social thermoregulation is almost always about warming up the body, not cooling it down. It is heat, not cold, that dwarf caimans steal from termites. Harry the Penguin huddles with his mates to keep snug, and so do naked mole-rats. There are a few instances in which animals rely on others to cool down. Camels, for example, may lie down in a group to help shade each other and to reduce the amount of surface area exposed to solar radiation. But these scenarios are the exceptions, not the rule.

There is a good reason for why social thermoregulation is mostly about keeping warm together. Heat can kill very quickly. Blood flow to the brain decreases, tissues get damaged. It's unlikely that groupmates could do anything quickly enough to save one of their number from temperatures that are too high. With cold, it's a different story. Unlike heat, cold is not a fast killer. When the mercury descends, an organism has at least some time to prepare, to seek the help of others.

These differences in how animals deal with cooling down versus warming up are also reflected in our own human bodies. From the way our brain's hypothalamus—in effect, the structure that coordinates our bodily thermostatic systems—is organized to the distribution of temperature receptors in our skin, we are each hardwired to see others as a potential source of warmth, both physical and, of necessity, psychological. This is where Chapter 4 takes us.

## CHAPTER 4

# People Are Penguins, Too

## *The Workings of Our Inner Thermostat(s)*

**Who doesn't like penguins?** They remind us of ourselves. With the possible exception of Charlie Chaplin, no creature has ever presented a more effective and affecting caricature of the human condition than the emperor penguin. As Chapter 3 suggests, penguins and people have much in common, including some disturbing practices we did not touch upon in that chapter, such as rape, pedophilia, necrophilia, and suicide. The British Antarctic explorer and naturalist George Murray Levick was a member of the 1910–1913 Scott Antarctic expedition and spent the austral summer of 1911–1912 at Cape Adare, the northeasternmost peninsula of Victoria Land (East Antarctica), observing the world's largest Adélie penguin rookery. He subsequently published a book, *Antarctic Penguins*, but the four pages of notes he kept on Adélie sexual behavior were considered too indecent for publication and lay hidden for a century. Levick himself found his own notes so shocking that he wrote them in classical Greek, just to keep them from ignorant eyes.

It was not until 2012 that the Cambridge University scientific journal *Polar Record* published the observations, which, in the words of the abstract, comment extensively "on seemingly aberrant behaviour of young unpaired males and females including necrophilia, sexual coercion, [and] sexual and physical abuse of chicks."[1] Levick did not note suicidal behavior, but others have, including the German film-maker Werner Herzog in his 2007 documentary film on people and places in Antarctica, *Encounters at the End of the World*, which includes footage of a penguin walking away from the others, march-ing off to certain death in the Antarctic interior.[2]

Narrated by Morgan Freeman, director Luc Jacquet's *March of the Penguins* was a big 2005 hit in movie theaters worldwide. The documentary followed penguins in their 70-mile journey in −70°F (−56.7°C) temperatures to a breeding ground where they huddle and protect their eggs. As the *New York Times* reported in a story published on September 13, 2005, many conservative Christian groups hailed the film as a celebration of the beauty and sanctity of life, evidence of intelligent design (versus natural selection), and an exultation of monogamy.[3] They would probably feel different if they perused the observations of Levick unearthed in 2012 or if they would simply try googling "Levick penguins." Online, people of all religious stripes will find penguins excoriated as "depraved," "jerks," and, yes, "assholes."

Have your laugh, but then dig a little deeper. What is disturbing about the behavior Levick and Herzog documented is that it is all, in a word, "antisocial." More importantly, it is evidence of antisocial behavior in animals that are mostly overwhelmingly social—just like us humans. Levick's notes on penguin sex reveal as much about Edwardian-era moral hypocrisy as they do about penguins, but Her-zog's image of a lone penguin willfully waddling off to his doom is genuinely affecting. Suicide, after all, is the ultimate renunciation of social behavior, and the penguin's means of self-destruction is the most literal abandonment of society, the company of others who, as we saw in Chapter 3, are the source of life-giving warmth. The penguin huddle is a social act, an act of social thermoregulation, a dramatic demonstration of the economy of action—in this case, the leveraging of penguin "society" to ensure sufficiency of energy resources for the preservation of life itself. To watch an animal, especially one that reminds us of our own species, walk away from all of this is profoundly unnerving, sad, and even tragic.

## Shivering brothers from other mothers

It is not just that penguins bear a comic resemblance to us—or, perhaps, we to them—it is that we share very basic mechanisms and strategies for regulating our body temperature, an activity essential both to optimizing our performance as an organism and, ultimately, to our survival. Like the Antarctic birds, we shiver in the cold, the veins in our skin constrict, and we use both our white and brown fat stores for energy deposits. Furthermore, just like the huddling penguins, we have also relied on others to keep us warm as we've evolved. In pretechnological human society, that reliance quite closely resembled penguin behavior by making use of variations on huddling. Today, central heating obviates the need for collective embraces just to keep warm, but you probably didn't build your home's central-heating system yourself, from scratch, and if it goes on the blink, you will likely get on the phone and call an HVAC expert for help. The quest for thermoregulation is a driver of social behavior in penguins and human beings alike.

Shivering is a thermogenic—that is, heat-producing—mechanism. Naturalists describe shivering as a "facultative" thermogenic mechanism, meaning that it is not the only mechanism animals use to generate heat. That said, shivering is the only universal facultative thermogenic mechanism in vertebrates that are endothermic—that is, both capable of and dependent on internally generating heat (as opposed to ectothermic animals, which depend on external sources to regulate their body temperature). Since skeletal muscle makes up a large part of body mass and accounts for a high scope of metabolic rate, the active contraction of muscles demands considerable energy and thus generates a significant amount of heat. Cold conditions induce the involuntary muscle contraction we call shivering.

Because muscle tissue has a large metabolic scope, its energy expenditure and, therefore, its heat production from a resting to an active state is greater than that of most other tissues. As the Finnish animal physiologist Esa Hohtola has concluded, evolution "selected" shivering as the main thermogenic mechanism in both endothermic vertebrate groups, birds and mammals.[4]

Yet it is also true that shivering is found in some heterothermic species. Some species vary between endothermic (so-called warm-blooded) and ectothermic (so-called cold-blooded) thermoregulation and are therefore called heterothermic. Such heterotherms as pythons (when brooding), moths, honeybees (when they warm themselves up for flight), and tuna all use muscular mechanisms of heat production analogous to shivering to warm up or to defend against heat loss. This allows them to attain partial endothermy.

In evolutionary terms, there is phylogenetic evidence that mammals and birds evolved from reptilian ancestors. For those reptilian ancestors, another term has been introduced into the thermoregulation lingo: gigantothermy. *Gigantothermy* captures the thermoregulation of very large, bulky ectotherms, their size allowing them to maintain a relatively high body temperature more readily than smaller ectotherms. Larger animals have proportionately less of their body close to the surface—and therefore close to ambient external temperatures—than smaller animals of similar shape. They are thus more fully insulated from the thermal whims of the outside world than small ectotherms. Based on this phenomenon, many evolutionary scientists suggest that theropod dinosaurs, the ectothermic ancestors of modern birds, were heterotherms, their large size making them capable of a high degree of endothermy. Nevertheless—and this is significant—birds and mammals evolved from *separate* reptilian ancestors, and no evidence exists of a common ectothermic, heterothermic, or endothermic ancestor for both birds and mammals. This suggests that shivering thermogenesis evolved in the two animal groups independently.[5] That is important, because there are key differences between how birds and mammals—and penguins and people—shiver.

In birds as well as mammals, shivering starts small, with the recruitment of small motor units. In birds, this progresses to the recruitment of larger units, which produce a crescendo of increasing shivering. In mammals, the initial recruitment of small units produces a so-called thermoregulatory muscle tone, which is followed by *grouped* discharges of motor units. To an external observer, the shivering of birds is virtually invisible because all the motion is internal. In contrast, an outside observer can clearly see mammals, people included, shivering. This "true shivering" is the result of higher contraction intensities produced by grouped discharges of motor units. Looking at this difference from the perspective of natural selection, we can conclude that the shivering mechanism of birds is well adapted to species that fly.

The gross movements of true shivering would make for a very turbulent flight or might even be altogether incompatible with controlled flight. Moreover, birds probably benefit from the lower intensity of their shivering tremors by suffering less loss of body heat through convection—heat transfer due to increased molecular movement within body fluids. In fact, small birds are more resistant to cold than comparably small mammals. Mammals, however, need not worry too much about the effects of intense shivering. They don't fly, and their shivering, while plainly visible, is not so extreme as to threaten locomotion or balance while on the ground. It is true that shivering may negatively impact muscle function in the cold, which might make it harder for a person to shovel snow or do other work efficiently. But human beings adaptively use distal muscles (situated farther from the body's center, farther out on the limbs) for work, and proximal muscles (nearer to the center of the body, the "midline") for shivering, thereby minimizing such problems. There is another difference between birds and mammals in their motor control of shivering. In both, the intensity of shivering is modulated by the respiratory cycle; however, in mammals, shivering is facilitated during the inhalation of breath, whereas in birds it is facilitated during exhalation.

For all the thermogenic help shivering gives to both penguins and people, it is a very expensive means of thermoregulation in terms of energy use. The more energy we use, the more fuel (food) we need. The more food we need, the more active we must be to get it, which, of course, requires more energy. It is a reactive cycle in which we are obliged to respond to the environment immediately. Adjusting to outside conditions *all the time* is tiring. Imagine having to give undivided attention to the environment every single second and changing your behavior accordingly. It is simply not feasible. It makes sense, therefore, to see shivering as the body's last resort for such reactive temperature homeostasis. This said, experimental evidence suggests that the thermal threshold needed for shivering to commence is lower than that of any other known thermoregulatory mechanism, but, from an energy perspective, its efficiency is only about 10 to 20 percent. Although effective, shivering is also very costly. You don't have to be a physiologist or experimental psychologist to reach this conclusion. Shivering is uncomfortable. It never feels "normal"— that is, anything like a sustainable condition. It is, in fact, fatiguing.

When you find yourself somewhere cold enough to start shivering, you feel a compelling impulse to seek an external source of warmth. In fact, being cold enough to shiver reveals to you the inadequacy of solitary attempts at thermoregulation. Lacking a fire or the means to quickly make a fire, lacking additional clothing or a blanket, you seek social thermoregulation. As with penguins, this may prompt some form of huddling. Or it may simply drive you indoors.

# BAT

The demands of thermoregulation are highly dynamic, driving organisms to mount a variety of defenses against life-threatening changes in temperature. Most of these, including shivering, as we've learned, are reactive and require a good deal of energy. Others are less demanding. Two additional means of thermoregulation, which human beings share with penguins, are vasoconstriction and the use of brown adipose tissue, or BAT.

Within what is called the thermoneutral zone (TNZ), people and other endotherms can maintain normal body temperature without increasing their heat production (either through an activity such as shivering or from an external heat source). The TNZ concept was introduced in 1902 by the German physiologist Max Rubner, a pioneer in metabolic research. The concept was applied specifically to humans in 1937 by James D. Hardy and Eugene F. DuBois in their paper "Regulation of Heat Loss from the Human Body."[6] Hardy and DuBois defined the lower extreme of the TNZ as "the maximum gradient ($T_{skin}$—$T_{air}$) over which the body can maintain its temperature without increase in heat production." In nude subjects, Hardy and DuBois found a maximum gradient of 7.38 Fahrenheit degrees (4.7 Celsius degrees) at the lower limit of the TNZ, which corresponded to an air temperature of 83.3° F (28.5°C). Below this limit, vasoconstriction of blood vessels in the skin was no longer sufficient to maintain body temperature.[7]

The blood that circulates through vessels in the skin, close to the surface of the body, acts as a heat-exchange medium, much like refrigerant circulating in the coils of a refrigerator or an air conditioner. When the air temperature cools, the surface vessels constrict, limiting the circulation of blood near the skin's surface, thereby reducing the amount of body heat lost to the air. Below the lower limit of the TNZ, vasoconstriction alone cannot sufficiently reduce heat loss to maintain body temperature. As is apparent, if you are naked, you have a very limited capability of maintaining body temperature unless you live in a very warm climate indeed. In most geographical locations, you need external help, beginning (usually) with clothing, which acts to keep the microclimate around the body within a more comfortable range. The colder the air temperature, the more clothing required, but, of course, there are only so many layers you can put on to maintain body temperature in very cold conditions over a period of time.

We humans do have some other resources for internally maintaining body temperature. We evolved as homeothermic endotherms, animals with both the necessity and the capability to maintain our body temperature at relatively high levels. This gives us the ability to very quickly adapt to a very wide variety of environments, a talent rare in the animal kingdom. And we are better at it than other animals, which means that our level of activity can be much higher than that of other mammals.

Our special talent comes at a cost, however. We need a lot of energy, which requires a lot of fuel. For us, thermogenesis, the production of heat, is fueled by the combustion of carbohydrate or fatty acids. Also essential to maintaining our adaptability to extreme environmental demands is, it turns out, sleep. When the U.S. writer Clement Clarke Moore composed his much-loved evocation of the "night before Christmas" ("A Visit from St. Nicholas," 1823), he was not just being poetic in his description of going to bed that night when he reported that "Mamma in her 'kerchief, and I in my cap, / Had just settled our brains for a long winter's nap." He was describing a fact of human physiology: lower temperatures cause us to sleep more. Our winter naps tend to be longer than our summer ones. Lower seasonal temperatures create greater energy demands, and we need to conserve our resources by lowering metabolism in sleep.

We humans have evolved morphological and physiological traits that are very different from those of many other mammals. Our high level of activity both demands that we secure food (sources of carbohydrate and fatty fuel) and enables us to do so. Sleep, which increases with lower temperatures, facilitates conservation of metabolic energy resources. Yet another evolutionary adaptation, the development of brown adipose tissue (BAT) stores, increases our nonshivering thermogenesis.

BAT is found in almost all mammals and seems to be especially plentiful in hibernating mammals (including rodents). For a long time, it was thought that in humans, only infants had BAT-deposit regions. It is true that BAT is abundant in human newborns—who isn't familiar with "baby fat"? In 2003, Christian Cohade and other researchers, using positron emission tomography (PET) scans, discovered BAT deposits in adult humans.[8] Since then, it has become commonly accepted that BAT not only persists in human adults but is thought to play a role in thermoregulation, producing heat by nonshivering thermogenesis (although exactly how much is still unknown). This role very likely emerged after our species lost its fur and was an adaptation that helped us cope with the cold winter nights during sleep in the equatorial savannah, where our species evolved.

In the economics of metabolism, there is no free lunch. Indeed, BAT costs us a considerable amount of energy when it is activated, typically in the form of lipid acids. BAT is thought to contribute up to 15 percent of total energy expenditure in human beings when it is in use. This said, because BAT content varies in human body weight,[9] I believe that it is also likely an important factor in *individual* differences in metabolism and people's social behaviors.[10]

We will have more to say about the heat-producing function of BAT later, but for now let's note that thermogenesis by BAT activity has been well described in rodents as a regulation mechanism for cold-induced thermogenesis. That is, as heat-producing tissue, BAT is activated by exposure to cold, not just in rats but also in humans. (People are rats, too!) Importantly, BAT is not the only component of the human adipose organ, which consists mainly of white adipose tissue (WAT), or white fat. WAT is used for energy storage and, making up as much 20 percent of the body weight in the human male and 25 percent in the female, when subcutaneous, acts as a thermal insulator. Unlike BAT, however, WAT is not recruited for thermogenesis. Moreover, WAT contains fewer capillaries than BAT. More generously furnished with capillaries, BAT distributes the heat it produces throughout the body.

# The dismal science of temperature

The Victorian historian and essayist Thomas Carlyle complained so frequently about his indigestion that many readers believe his well-known pessimistic gloom can be attributed to a chronic bellyache. But when he bestowed upon the study of economics the label by which it is still known—"the dismal science"—it wasn't just his gut talking. Carlyle was acknowledging the assumption at the core of all economic theory: scarcity. All economists assume a gap between limited ("scarce") resources and theoretically unlimited human wants and needs. On a practical level, economics is therefore all about trying—somehow—to manage the dismal gap efficiently and effectively.

Organisms—penguins, humans, every living thing—must successfully manage an analogous gap in the energy economy. The sources of energy are limited. The demand for energy is theoretically without limit. Ectotherms depend on external sources of heat energy and therefore are not obliged to generate their own heat internally. But the dismal science of energy economics dictates that ectotherms must sacrifice the opportunity for a greater range of activity and environmental flexibility. Endotherms maintain body temperature largely through the internal means of generating heat we have just discussed. This ability buys them a greater range of activity and the potential to adapt to a greater range of environments. The dismal cost to the endotherm? Spending a great deal of time and energy looking for sources of combustible fuel (food) to feed the furnace of metabolism. Moreover, if the internal systems of an endotherm fail to maintain body temperature within a narrow window, performance, health, and even survival are put in jeopardy.

All organisms must somehow arrive at a workable economy of action. Those that succeed in sufficiently solving their economic problems survive in sufficient numbers to reproduce and thereby win the lottery of natural selection. Those that fail suffer extinction. Naturalists have noted that, in many animal species, working the equation of the economy of action involves a combination of internal and external resources. Penguins, for instance, are endotherms, but they maintain body temperature through a combination of internal metabolic thermogenesis and external energy sources, especially huddling. That is, they practice social thermoregulation. Despite their range of internal thermogenic and thermoregulatory systems, they have not developed social thermoregulation as highly as human beings—no species has. Human biological evolution selected for sophisticated internal systems, which both enable and require high levels of activity to obtain the necessary sources of metabolic fuel. These levels of activity, in turn, both enable and require social evolution. Thus, we may think of the economy of action as driving social thermoregulation, which drives a wide variety of human social behaviors. These, in turn, play essential roles in the evolution of culture and what we call society and civilization.

One of the paradoxes of the concept of the economy of action is that so many professors of economics use the concept of *Homo economicus*—economic man, or human—as a label for a fictive species of humans whose representatives are consistently both rational and narrowly self-interested, forever calculating costs versus benefits in terms of themselves, the individual. In fact, as a pedagogical construct, *Homo economicus* is flawed. It is not only inadequate to account for much that happens in economics but also reflects a fundamental misunderstanding of how human beings have evolved to analyze costs versus benefits in the economy of action. Our brains and bodies have evolved to base cost-benefit analysis not on ourselves, as individuals in isolation, but instead in social terms. We ground our decisions in the expectation that others will be there for us, something my good friend and colleague Jim Coan has called our "social baseline."[11] The penguin predicts the weather in terms of social capital, the expectation that others will be available for huddling. We humans likewise thermoregulate socially.

In their provocative picture of culture and civilization, *Connected*, Nicholas A. Christakis and James H. Fowler invented their own variation on *Homo sapiens* to more accurately reflect the evolutionary status of contemporary humanity. For them it is not *Homo economicus*, but *Homo dictyous*—"network man" or "networking human."[12] In terms of Coan's social baseline theory, humans add a social-capital contribution to their portfolio of energy investments, which pays dividends from others. *This* is social thermoregulation, and it depends on a cognitive faculty that is very highly developed in humans.

Many organisms are equipped with temperature-detection organs, which function as part of a monitoring and response system, enabling them to respond appropriately to changes in temperature. Humans share with other animals the ability to sense and respond to temperature. Probably unique to humans, however, is the extent to which we have developed an additional ability. Not only can we detect and *respond to* change, but we can also readily predict changes in ambient temperature long before such changes might impact core body temperature. Humans can thus act proactively and even preemptively in anticipation of likely changes (much more so than other animals, although many of them also anticipate the seasons by migrating). This predictive faculty impacts social thermoregulation in people and, thus, the nature of society and culture. Partly in anticipation of temperature changes, people create clothing, build housing and communities, and develop technology. Indeed, thanks to the elaborately developed human predictive faculty, the detection of temperature involves predictive systems outside of our physical bodies. Culture affords us the development and use of technology to detect temperature and to predict changes many days in advance.

That same level of technology enables ambient temperature control through heating and cooling systems, as well as through thermoregulatory clothing such as coats, shorts, socks, and the like. While the area encompassing all latitudes between the subtropics and the polar circles is called the Temperate Zone, it should more accurately be labeled something like the Incredibly Variable Zone, because climate, weather, and temperature vary so widely in this huge geographical swath. Those who live in the misnamed Temperate Zone become quite skilled at predicting the weather and selecting clothing accordingly. Dissatisfied with our well-developed seat-of-the-pants approach, scientists recognize that the insulative quality of a given article of clothing depends on such factors as its dryness resistance, evaporative resistance, and compressibility from wind and movement; accordingly, they've compiled exhaustive databases on the thermal properties of clothing. They have developed the "clo" as a unit of measurement to mathematically express insulation value:

$$1 \text{ clo} = 0.155 \text{ K·m}^2\text{·W}^{-1} \approx 0.88 \text{ R (where R means ft}^2\text{·°F·hr/Btu)}$$

In real-world terms, if you wear walking shorts with a short-sleeved shirt, you wear the insulation benefit of $0.36_{cl}$ (little more than a third of a clo). Put insulated coveralls over long-sleeved thermal underwear and long-underwear bottoms, and you get $1.37_{cl}$. Clothing of $1_{cl}$ lowers the ambient temperature required to maintain comfort by 17.46 Fahrenheit degrees (9.7 Celsius degrees).

Is it an absolute necessity for all of us to do the math before we leave the house in the morning? Not really. Embodied cognition, combined with experience, and aided by weather reports, gives us the practical ability to predict and control our personal weather by dressing "appropriately."

Technology allows us humans to predictively outsource mental functions to the environment. The ability to prepare for predicted changes in ambient temperature derives from cognitive mechanisms altogether different from those related to more urgent, proximal thermoregulatory adjustments. We will detail these proximal mechanisms shortly—they involve a distributed thermostat system coordinated by the hypothalamus, which is located in our neural utility basement, deep within the forebrain. In contrast, the predictive cognitive mechanisms are located in "higher"-function brain areas, the prefrontal cortex and, below this, the cingulate cortex.[13] These cortical structures are home to mechanisms related to vigilance, working memory, and executive control. By predicting temperature in advance, the cognitive processes in these brain areas preclude the necessity of downstream thermoregulatory efforts. In effect, the downstream processes are outsourced to technology so that energy expenditure in service to thermoregulation can be organized and carried out more efficiently. Armed with the technological means of temperature detection, we do not need to rely on more proximal means of detection, such as the feeling of temperature on the skin. By the time our bodily sensors tell us that it is getting uncomfortably or even dangerously hot or cold, it can be too costly to start taking the steps to make the necessary changes. Predicting and planning avoid the necessity of resorting to more bioenergetically expensive responses to change.

If you are a fan of the ancient Greek fabulist Aesop or his seventeenth-century French translator, Jean de La Fontaine, you remember the fable called "The Ant and the Grasshopper." The ant is a prudent and industrious plodder, who diligently lays up stores of food against what he "predicts" as the coming of winter. The grasshopper, in contrast, spends the summer singing and is hard-pressed indeed when winter comes. When the grasshopper, shocked to find himself dying of hunger, begs the ant for food, the little fellow coldly tells him to dance the winter away. If you prefer something more modern, think about two iconic U.S. investors. Warren Buffett became the fourth-richest man in the world through an industrious, plodding, and prudent program of investment. In terms of his personal conduct and lifestyle, he has no interest in the fast track and lives a very low-key life in Omaha, Nebraska. CNBC reports that he never spends more than $3.17 on breakfast, lives in the same house he bought for $31,500 in 1958, and drives a 2014 Cadillac XTS, a nice $45,000 car but certainly not a Bentley. Buffett lives with an eye toward the future rather than a desire to revel in the present. His antitype is Jordan Belfort, the infamous "Wolf of Wall Street" played by Leonardo DiCaprio in the 2013 Martin Scorsese movie. Belfort made a staggering fortune very quickly in the 1990s with his

Stratton Oakmont brokerage firm, using it to fuel an orgiastic Wall Streeter lifestyle that would make Caligula blush—only to go bust (and to prison) because both his lifestyle and his investment style were unsustainable.

Warren Buffett worked—and, at 89, continues to work—hard, but he does not have to expend vast amounts of energy feeding a frenetic lifestyle while desperately trying to dodge the law, which was Belfort's fate. As predictive animals, we human beings are very active in our efforts to provide for what we predict as our future thermoregulatory energy needs. It's hard work, but it gives us an advantage over organisms that are unable to predict the weather— both actual and social—and can respond only reactively to changes in thermal conditions. In the short term, their energy demands can be very high. They shiver, vasoconstrict, and draw on BAT energy reserves, and when none of these things work, they may hibernate or enter a state of torpor. Some have the ability to huddle. If all of these strategies prove inadequate to sustain thermoregulation within a viable range, they die.

But let's pause here before our analogies run away from us. Warren Buffett's methodical investment style has been spectacularly successful for him; however, he understands that the careful analysis and deliberate planning behind his predictions do not excuse him from reacting, when necessary, to sudden drastic changes in the investment landscape. Sometimes he must pounce on an opportunity as if there were no tomorrow, and at other times he must quickly shed a bad investment. Likewise, we need to remind ourselves that the ability to predict temperature change and the ability to do something about it, important as they are, by no means relieve humans of the need for the level of proximal detection that drives reactive thermoregulatory responses. Our bodies are equipped with thermosensitive neurons capable of detecting a temperature range from the noxious and dangerous—greater than 125.6°F (52°C)—to the innocuous, approximately 71.6°F to 104°F (22°C to 40°C).

Although "normal" core body temperatures vary by individual, the traditional normal value for human oral temperature is 98.6°F (37°C). However, the normal human core body temperature undergoes a regular fluctuation of 0.9 to 1.8 Fahrenheit degrees (0.5 to 1.0 Celsius degrees) in the course of the daily 24-hour (or "circadian") cycle. Indeed, core temperature depends more on time of day than on activity. It is normally lowest during sleep, is slightly higher in an awake, relaxed state, and rises slightly higher still with activity. Significantly, oral temperature is also influenced by social factors. As we will see later, the diversity of your social network and your feeling of social connection or disconnection—think *Homo dictyous*—have a measurable impact on oral temperature.

Physiologists and experimental psychologists often find it useful to consider the human body as an inner core within a peripheral shell. The inner-core temperature normally is maintained at about 98.6°F (37°C), while the outer-shell temperature depends on environmental conditions (such as ambient temperature) and vasomotor tone, which is a blood vessel's degree of constriction relative to its maximally dilated state. Our outer shell's thermosensitive neurons are divided into cold receptors, which react in the temperature range of approximately 23°F to 109.4°F (−5°C to 43°C), and warm receptors, which operate only at temperatures greater than 86°F (30°C). Note that cold receptors are about 10 times more abundant than warm receptors. Monitoring peripheral temperatures, temperatures at the outer shell, is generally concerned with detecting cold rather than warm temperatures. That is, there is a marked asymmetry between how we detect heat and cold.

Because our skin (outer shell) has many more cold receptors than warm receptors, it is therefore set up primarily to detect decreases in ambient temperature. In the brain (inner core), we have more heat receptors than cold receptors, and thus the brain is built to detect increases in inner-core temperature rather than decreases. Temperature increases are more immediately dangerous than decreases. Detection of an increase requires a rapid response—as in, for example, drawing back reflexively from a source of intense radiant heat, such as a hot stove. We must down-regulate immediately. In contrast, responding to a temperature decrease, up-regulating, is less dangerous and can be done far less urgently.

## A single inner thermostat?

What are the brain mechanisms that control thermoregulation, and in humans how do these allow for immediate, urgent, and reactive responses as well as longer-term predictive action? Let's begin to answer these questions by going back to 1878, the year in which the French pioneer of physiology Claude Bernard died and his final work, *Leçons sur les phénomènes de la vie communs aux animaux et aux végétaux* (*Phenomena of Life Common to Animals and Plants*), was published.[14] In it, he wrote: "The stability of the internal environment" of an organism—the *milieu intérieur*—"is the condition for the free and independent life." He went on to explain that the "living body" needs the environment that surrounds it but is "nevertheless relatively independent of it." It achieves this state because its tissues are "withdrawn from direct external influences and are protected by a veritable internal environment which is constituted, in particular, by the fluids circulating in the body." Variations in the external environment are "at every instant compensated and brought into balance" by the interior environment, so that "far from being indifferent to the external world, the higher animal is . . . in a close and wise relation with it," creating an equilibrium through "a continuous and delicate compensation." In 1932, the U.S. physiologist Walter Bradford Cannon gave this internal-external balancing act the name by which it is known today: "homeostasis."[15]

Cannon argued that much of what an organism does is in direct service to maintaining homeostasis. With respect to thermoregulation, Cannon was mainly concerned with reflexive mechanisms in mammals and discussed active physical means of heat production and heat loss, such as shivering and vasomotor changes, as well as chemical means of heat production, as through increased thyroid and adrenal output. So intellectually compelling was the combination of Bernard and Cannon that, for many years, homeostasis was the default assumption underlying psychological theories of drive reduction and was used rather indiscriminately to explain far more complex behaviors than those Cannon himself had studied.

In 1940, in an effort to locate and investigate brain structures involved in maintaining homeostasis, the U.S. neurophysiologist Stephen Ranson pioneered methods of creating electrolytic lesions (destruction of neural tissue by electrical shock) with great precision by inserting electrical probes into the brain using a three-dimensional surgical procedure. Ranson applied the technique to cat and monkey brains and found that creating the destructive lesions in the preoptic area (POA) of the anterior (front) hypothalamus rendered the animals incapable of keeping their body temperature from rising in the heat. The animals retained their ability, however, to maintain near-normal body temperature in the cold. When Ranson instead created lesions in the posterior (rear) hypothalamus, he found that his cats and monkeys were unable to regulate body temperature in both heat and cold.

Ranson theorized from these results that the preoptic, or anterior, hypothalamus controlled the down-regulation of body temperature, but because it also sent outgoing nerves into the posterior hypothalamus, when this area was destroyed, the ability to down-regulate and up-regulate—to regulate temperature at all—was lost. Put another way, Ranson concluded that what he called the brain's "heat-loss center" was located in the POA of the anterior hypothalamus. What he labeled the "heat-production center" was in the posterior hypothalamus. The clear implication of a single structure with one area down-regulating body temperature and the other up-regulating it was that the hypothalamus functioned as the body's thermostat; and so far as thermoregulation goes, it was the brain structure central to monitoring and maintaining homeostasis.[16]

Ranson's result was important. Unfortunately, however, the neat anterior-posterior division of hypothalamic thermoregulatory function was not supported by the totality of Ranson's own carefully acquired data. When he created symmetrical lesions in the rostral (anterior, or forward-facing) hypothalamus in cat brains, the animal was no longer capable of adequately maintaining body temperature in the cold. This result clearly contradicted the theory that the anterior hypothalamus was that structure's "heat-loss center." If it were, even a cat whose POA had been destroyed but whose hypothalamus was otherwise intact should have no trouble maintaining body temperature in the cold.

People may be like penguins, but even scientists are human, too—and sometimes all too human. As professor of neurology and director of the Institute of Neurology at Northwestern University, Ranson was so highly esteemed a researcher that no one immediately questioned the conflict between his data and his conclusion. Ranson himself died suddenly, at age 62, of coronary thrombosis just two years after his experiments on cats and monkeys. It was years more before studies began to be published confirming that lesions created in the POA not only of cats, but also of goats and rats, rendered the surgical subjects incapable of adequately up-regulating body temperature in the cold.

As the late University of Delaware physiological psychologist Evelyn Satinoff noted, neither Ranson's studies, nor those on rats, cats, and goats that eventually followed, attempted to answer the question of whether the POA is, in fact, a central thermostat. Does it alone sense relevant temperatures and transmit the information to controllers that may be located elsewhere in the nervous system? Ranson's and other studies demonstrated nothing beyond that an intact POA was necessary to maintain normal body temperature. Nevertheless, several studies in the twentieth century have been consistent with the idea that the POA of the hypothalamus does function analogously to a thermostat. The most suggestive of these was a 1964 experiment Evelyn Satinoff performed on rats that had been conditioned to press a bar to turn on a heat lamp. When the rostral (anterior) hypothalamus of these animals was cooled, they pressed the bar much more frequently than when it was not cooled.[17] Right around the same time, another researcher, Harry J. Carlisle, heated the same area of the hypothalamus and elicited the opposite effect. The rats' response to cold conditions—obtaining warm air by pressing the bar—was depressed.[18] Satinoff's and Carlisle's results strongly suggested that motivated nonreflexive behavior could be influenced by varying the temperature of the POA. This added to the experimental evidence that the anterior hypothalamus is a thermostat.

## Toward an embodied cognition of social thermoregulation

The concept of a single thermostat was simple, and the experimental data accumulated by the end of the 1960s seemed quite compelling. Added to this was ample evidence that, whether or not the hypothalamus functioned as *the* body's thermostat, or *the* pilot in the ship, the structure was necessary for adequate thermoregulation. On top of that, the theory that the hypothalamus was a relatively simple thermostat accorded neatly with the universally accepted notion that homeostasis was the ultimate objective of thermoregulation as well as other regulatory processes. Moreover, picturing the hypothalamus as a mechanistic negative-feedback control did not require attributing any complex cognition to the thermoregulatory process. Like an electromechanical thermostat, the hypothalamus was "designed" to detect "errors" between a reference setting and a feedback signal, initiating the appropriate action in response. As a household HVAC thermostat activated the furnace or air conditioner in response to a detected variation from a set temperature, so the hypothalamus switched on or off reflexes and behaviors to up- or down-regulate body temperature as required.

Thinking of the hypothalamus as a simple thermostat brings us all the way back to the universe of Cartesian mind-body dualism.

Maybe the whole brain was not a pilot in a ship—controlling the vessel but not itself a part of it—but *this* part of the brain, the hypothalamus, must be such a pilot. As the thermostat on the living-room wall controls the operation of the furnace down in the basement but is not a part of it, so the hypothalamus was seen as controlling an organism to which it was "wired" but of which it was not organically a part.

Evidence once again stubbornly emerged to make this mechanistic pilot-in-the-ship interpretation inadequate. In 1970, Evelyn Satinoff and Joel Rutstein conducted an experiment with rats that were trained to use a heat lamp to warm themselves when temperatures dropped. The researchers made surgical lesions in the preoptic area of the hypothalamus (POA) of some of the trained rats. These animals and a control group (which had been trained but not operated on) were exposed to cold. When body temperatures dropped as much as 11.7 Fahrenheit degrees (6.5 Celsius degrees) after an hour of exposure, both the control and the lesion rats were still able to maintain their body temperatures within 1.36 Fahrenheit degrees (0.75 Celsius degrees) of normal for two hours by pressing a bar that turned on a heat lamp. Satinoff and Rutstein concluded from this that enough thermosensitive cells and neurons existed *outside* of the POA to enable rats to act to relieve the discomfort of cold and maintain nearly normal body temperatures even when POA lesions had disabled their ability to respond immediately. The trained rats' behavior was a *cognitive* compensation for what their disabled reflexive system could no longer do.[19] Additional experiments by Satinoff and others between 1968 and 1971 provided further evidence that *behavioral* thermoregulation survived destruction of the POA. The inference was clear. There were separate neural networks for behavioral and autonomic/reflexive thermoregulatory responses.

Unlike a totally reflexive response, the behavioral response does not support the Cartesian pilot-in-the-ship thesis. It is a cognitive response. Studies completed during the 1970s demonstrated the existence of many temperature-sensitive neurons behind the POA, in the posterior hypothalamus as well as in other structures, including the midbrain reticular formation and, below that, the medulla oblongata. Combining this finding with the persistence of behavioral thermoregulation in conditioned rats with large preoptic lesions, we must ask if thermoregulation is controlled by something more than a single thermostat.

But what about homeostasis? The idea of a simple "error"-correcting thermostat is ideally suited to what is widely considered the homeostatic necessity of maintaining a "normal" body temperature within 1.8 to 3.6 Fahrenheit degrees (1 to 2 Celsius degrees). Indeed, most mammals do maintain body temperature within these tolerances; however, it is also true that some "primitive" mammals, including sloths, hedgehogs, opossums, and tenrecs, cannot manage so narrow a window. Furthermore, even advanced mammals experience wider variation than a degree or two during, for example, seasonal hibernation, situations of food scarcity, pregnancy, and emotional stress. Dig more deeply into the question of homeostasis and body temperature, and you discover that there is no universal agreement on just what variable temperature is regulated to maintain homeostasis. The unspoken assumption is that the target is deep-body or brain temperature. But physiologist Michel Cabanac, whose early research was devoted to thermoregulation, has argued that the regulated temperature variable in small animals is skin temperature, while in large animals, including humans, it is deep-body temperature.[20]

As Satinoff came to suggest, homeostasis is not in danger of being disproved. It is necessary to life. But the negative-feedback control associated with homeostasis is not a simple, narrow, virtually binary system explicable by analogy with an electromechanical thermostat, which recognizes but two temperature states, the right temperature (the thermostat setting) and any deviation from that. Instead, Satinoff suggested, homeostatic thermoregulation can actually be quite flexible and is capable of being influenced by localized and even trivial changes in external temperature. Apply heat to the scrotum, and deep-body temperature drops more than 3.6 Fahrenheit degrees (2 Celsius degrees), even without a change in ambient temperature.

None of this overturns the concept of homeostasis. It just renders it more elastic and, even more importantly, it tells us that assigning thermoregulation to a single, simple thermostat totally localized in the hypothalamus is not supported by the accumulated experimental data. Nor is it compatible with evolutionary considerations, into which we are about to dive.

## Thermoregulation: The Warren Buffett way

Taking the study data into account, it is reasonable to conclude, with Evelyn Satinoff and others, that mammals, including people, have more thermostats than thermoregulatory responses. We can reach this conclusion because mammals up-regulate and down-regulate in more ways than a singular electromechanical thermostat can account for.

The idea that the nervous system consists of more than nerves entering and exiting the brain is hardly new. In the 1870s, the English neurologist John Hughlings Jackson argued (quite controversially) for an evolutionary hierarchy to explain the structure of the nervous system.[21] The higher nervous centers, he theorized, evolved from lower nervous centers distributed along the neuraxis, the axis of the central nervous system. These centers, he noted, were each capable of independent action, but normally they operated in concert and were integrated hierarchically. Satinoff found support for this hierarchical integration by noting that rats whose POA had been surgically destroyed exhibited abnormally high metabolic rates and therefore abnormally high body temperatures in room-temperature environments. In the cold, however, rats with POA lesions showed abnormally low levels of shivering and lower-than-normal metabolic rates. At the same time, their vasoconstriction in cold conditions was normal. These apparent anomalies led Satinoff to conclude that all the separate integrators (or "nervous centers," as Jackson called them) are independent—yet, in animals with normal, intact brains, they are hierarchically controlled.

Satinoff cited numerous studies that support a Jacksonian hierarchical structuring of how the central nervous system controls thermoregulation. In one study, cats' spinal cords were surgically severed at about the middle of the spine (level T6). Cooling below the level of this cut caused shivering as well as hind limb vessel constriction, even though the surgical separation prevented any neural transmission to the brain. The removal of the cerebrum in cats also removed their ability to shiver in their front legs when their entire bodies were cooled but still permitted shivering in response to cooling of the spinal cord. If the surgery is performed in the same animal lower in the brain, shivering and vessel constriction in the front legs is reinstated. This result indicated the existence of a region in the midbrain and upper pons that inhibits some activity in lower brain regions. If the influence of the upper areas is removed by surgical incision that disconnects the upper from the lower areas, these areas can themselves produce thermoregulatory responses. This is yet another indication of the existence of "nervous centers" or integrators capable of acting independently from higher brain centers. Under normal conditions, when the connections are intact and operating as they should, these lower centers are not independent but rather are coordinated by centers at a higher level of the brain. The effects produced by these surgeries are consistent with a system in which components of the body's "thermostat" are distributed in a hierarchy along the axis of the central nervous system, or neuraxis. The results are not consistent with the hypothalamus-as-thermostat theory.[22]

A thermoregulatory thermostat made up of a hierarchy of control centers distributed across the neuraxis explains how basic reflexes can be incorporated into a variety of action patterns. Moving up the hierarchy, greater complexity is added to the action of lower mechanisms. Jackson put it this way in 1882: the "higher the [neural] centre the more numerous, different, and more complex, the more special movements it represents." That said, we are still dealing with reflexes—reactive mechanisms. When predictive elements are added, we get the social-weather report that is the product of truly embodied cognition. In other words, the Warren Buffett way of thermoregulating—or, behaving like *Homo dictyous*. It is the coordinating mechanism at the summit of the hierarchy, ensuring that appropriate reflexes as well as more complex behaviors are activated promptly while inappropriate ones are suppressed.

As coordinated by the hypothalamus, thermoregulatory input is the data on which the predictive work of cognition is performed. For Jackson, the neural hierarchy was not merely anatomical but truly the product of evolution. Beyond the neural integrator that is the hypothalamus are the higher levels of cognition, which are integrated into the hierarchy. They are the equivalent of the pilot *if* the pilot were, in fact, part of the ship. The evolutionary hierarchy represents a level of increasing embodied cognition when elements contributing to the cognitive process are distributed along the neuraxis rather than concentrated in a single structure. Where the drivers of thermoregulation prompt predictive behaviors based on social capital—when penguins seek other penguin bodies for warmth or when humans rely on the warmth of friends and loved ones—embodied cognition informs plans of adaptive behavior. Both penguins and people become investors in social capital in order to earn dividends paid in energy, which may enable the activity that is required to find the food necessary to fuel endothermic thermogenesis or that may be spent more directly on ensuring the presence of huddling partners. In humans, the imperatives of thermoregulation drive many more abstract patterns of social thought and emotion, of the desire to find the company of "warm" people and to behave in ways calculated to avoid being shunned and left out in the cold.

## An evolutionary crossroads

Like penguins, we are endothermic homeotherms, both *capable* of generating our own heat and *obliged* to do so, as well as both *capable* of maintaining our body temperature within a narrow range and being *obliged* to do so. We share many features with penguins and with other animals. Climb far enough back down the evolutionary tree, and you find more common branches, ever-thickening boughs, and, eventually, a common trunk. There are, of course, differences between us and penguins. Many of the differences between people and other animals are obvious at a glance, but as we have seen, some are subtler. In terms of what these subtle differences tell us about evolution, they are nevertheless quite significant, as we've learned. Nevertheless, penguins and people struggle to solve the same problem: maintaining a viable body temperature.

Perching on the summit of evolution can give us humans a big head. *You penguins are very cute, huddled all together like that, but haven't you heard of a fireplace, a radiator, or forced-air heating?* And if penguins are not as bright as people, what about the ectotherms, the creatures we used to call "cold-blooded"? They're so far down the evolutionary ladder that they can't even generate their own heat!

Well, let's get ourselves grounded. As we saw in Chapter 3, we endotherms may be nature's hard-charging investment bankers, always hustling to build our internal energy fortunes, but there is something to be said for nature's easy-going slackers, the ectotherms, who take from the environment whatever heat energy they can get. Yes, the slacker lifestyle limits their range of evolutionary and individual options, but it requires much less energy. Viewed from an anthropomorphic perspective, more activity is better than less activity, yet, biased toward our species as we are, we nevertheless must admit that being so very active is far more expensive in terms of energy required. Much of an endotherm's activity is devoted simply to finding fuel. If we can for a moment shed our anthropomorphic bias, we will see that endothermy and ectothermy are two approaches to the same economic problem, each with its own costs and benefits.

Economy of action sometimes requires what may seem like extreme measures—again, as viewed from an anthropomorphic perspective. Some mammals are *obligate* hibernators, annually obliged to enter a period of hibernation regardless of ambient temperature and availability of food. Some other mammals are *facultative* hibernators, entering hibernation only in response to cold or in situations of food scarcity. Properly speaking, ectotherms do not hibernate, but many do enter into dormancy when cold conditions or decreased availability of oxygen depresses metabolism. Before you decry the low level of evolutionary development that demands such a sacrifice of consciousness and activity in service to the economy of action, think about our own human necessity of sleeping a third of our lives away—and on a daily basis, to boot.

The level of activity both available to and required from higher homeothermic endotherms like us demands refractory periods of sleep or other forms of reduced consciousness and decreased metabolic activity. Homeothermic endothermy also requires a high level of complexity in the thermoregulatory system. As we have seen, it is far more complex than the set of feedback loops between thermosensitive neurons and a single thermostat, the hypothalamus. That traditional model was better suited to the Cartesian paradigm of a separate mind and a separate body. The whole progress of modern physiological psychology has been toward distributing more of cognition throughout the body and the social world, so that mind, body, and, crucially, other people are thoroughly integrated.

The thermostatic mechanism of thermoregulation, we now see, may be integrated in the hypothalamus but is not simply controlled by it. Rather, that structure is at the hierarchical summit of neural structures along the neuraxis. The thermosensitive neurons distributed throughout the body's shell differ from those closer to the body's core and perform different but coordinated functions. On both the autonomic and cognitive levels, the neurological systems associated with thermoregulation in humans are extensively embodied. Moreover, if we consider *other* people to be part of *our* psychology—which, as *Homo dictyous*, we must—these neurological systems are not just thoroughly *embodied* in one's own body but are also thoroughly *grounded* in other people.

Grounded cognition, as demonstrated in human thermoregulation, is a long, long way from the Cartesian pilot-in-the-ship model. In Chapter 1, we introduced the idea that penguin huddling is a social behavior driven by thermoregulation. Penguin huddling is a large-scale operation, not a last-minute emergency move. It requires each penguin to behave in a social way to the extent of surrounding itself with other penguins. Each bird is in the position of predicting its *future* body temperature based on its *current* social capital—the presence of other penguins. In effect, the penguin creates a weather report, a *social*-weather report, and a prediction of its own body temperature based on both the weather report and the social-weather report. How early penguins developed this predictive skill for the purposes of long-term thermoregulation we don't know. What we do know is that evolution selected for the trait of accurately predicting the presence and behavior of others. We can even reach this conclusion simply by noting which penguins survive in groups. Poor predictors are far less likely to huddle, to survive into adulthood, and to reproduce. Their genetic material disappears from the gene pool.

Humans, whose cognitive ability is evolved well beyond that of even the smartest penguins, developed more-sophisticated forms of weather prediction capable of driving social behavior. This behavior includes huddling but also more culturally evolved means of outsourcing the production of heat—everything from "discovering" and building fire, to finding and then building shelter, to ultimately developing increasingly sophisticated technology to reliably heat one's shelter. Operating through evolution, natural selection favors the survival and reproduction of those organisms best adapted to their environment.

Like other animals, we humans certainly react to the environment and do so continuously, in real time. But we also excel at what the emperor penguin does in only a rudimentary way: we predict the social weather. Instead of simply responding to the environment, which is a very costly and potentially dangerous strategy, we predict the future and plan for it. We possess the cognitive equipment to create both weather reports and social-weather reports, which proactively puts us in position to cope with whatever temperatures we encounter. *Biological* evolution provided the cognitive platform that enables us to predict and to plan, but *cultural* evolution extended and enlarged the scope and accuracy of prediction and multiplied the technological options for outsourcing such urgent yet long-term needs as thermoregulation outside of thermoneutral zone (TNZ) environments. This uniquely human adaptability to a wide range of habitat temperatures shows that our social interactions are not the conceptualized expressions of cognitive metaphor, as researchers George Lakoff and Mark Johnson argue, but are the products of mind distributed throughout the organism. We don't necessarily have to throw Descartes's pilot overboard; we just have to start thinking of him as one with his ship.

## CHAPTER 5

# Rat Mamas Are Hot

## *Temperature and Attachment*

The title of what many consider a "classic" study from 1983 says it all: "Autonomous Nervous System Activity Distinguishes among Emotions."[1] Using 12 professional actors, researchers Paul Ekman, Robert W. Levenson, and Wallace V. Friesen evoked emotion by two methods. In the first method, face posing, they instructed their actor-subjects, muscle by muscle, to make and hold for 10 seconds each a series of facial-emotion stereotypes, including anger, fear, sadness, and others. In the second, they asked their subjects to remember—to "relive" for 30 seconds—past experiences, each of which involved a specified emotion. To relive anger, a subject might recall a vivid insult. For sadness, the memory was perhaps of a loved one's passing. And so on. Each subject's facial movements were videotaped, and the researchers simultaneously recorded physiological measurements, including peripheral temperature (temperature measured at the extremities, such as a finger), heart rate, forearm tension, and skin conductance—also called "electrodermal response," when skin momentarily becomes a better conductor of electricity when exposed to physiologically arousing stimuli.

Ekman and his colleagues found that heart rate increased more during "relived" anger, fear, and sadness as compared to happiness, surprise, and disgust. Change in finger temperature was greater in anger (+0.18 Fahrenheit degrees [+0.10 Celsius degrees] on the left finger, +0.144 Fahrenheit degrees [+0.08 Celsius degrees] on the right) than in happiness (−0.126 Fahrenheit degrees [−0.07 Celsius degrees] on the left finger, and −0.054 Fahrenheit degrees [−0.03 Celsius degrees] on the right). They also found that, in directed facial posing, three subgroups of negative emotions could be distinguished on the basis of heart rate and finger temperature differences, namely, anger, fear, and sadness. When subjects were directed to make an angry expression, peripheral temperature changed nearly +0.27 Fahrenheit degrees (+0.15 Celsius degrees). Directed to make a face showing fear, temperature changed about −0.018 Fahrenheit degrees (−0.01 Celsius degrees). Making a sad expression caused a +0.018 Fahrenheit degree (+0.01 Celsius degree) change.

Based on their data, Ekman and colleagues concluded that their study provided the very first evidence that physiological changes could help distinguish four negative emotions (as well as positive versus negative emotions). More recent studies along the lines of Ekman's work include one from 2013. Stephanos Ioannou and colleagues asked children to play with a toy they were told was the experimenter's "favorite." They were not told that the toy was rigged to break. When the toy broke as the child played with it, the experimenters measured a marked drop in the child's peripheral temperature. When the experimenters soothed the child after the mishap, peripheral temperature rose, suggesting that the distress was relieved and perhaps even overcompensated.[2]

These are intriguing findings with regard to social thermoregulation, because specific emotions, evoked by directed facial posing or by acts of recall, are associated with specific peripheral temperature change (among other autonomic nervous system responses). In a general way, the findings reinforce some of the basic assumptions of embodied cognition. In fact, many emotion theorists would likely see in these findings support for William James's idea that bodily changes cause emotional experiences (we do not flee because we feel afraid; we feel afraid because we flee). Yet, as with so much in the

study of mind and body, merely monitoring certain measurable autonomic changes like temperature and heart rate does not fully explain
the multiplicity of factors and inputs involved in social thermoregulation any more than, say, vision is explained by measuring the electrical activity of photoreceptors in the human retina in response to
a flash of light. Vision is far more complex than what a measurement or two can tell us, and the realities involved in emotion theory
are more varied and complex than what Ekman and his colleagues
measured and analyzed. Almost certainly, increases and decreases
in peripheral temperature are not solely coupled to emotions—and
emotions are not the sole result of such changes. After all, our cognition is "grounded" in the social world, and temperature changes
must therefore be heavily dependent on the context in which an emotion is lived.

Cognition is grounded in a way that provides information about
the world, especially the highly social world in which we live. Emotions give us signals, which means that the changes in peripheral
temperature are also signals, ones that aid us in predicting behaviors of those around us who may help, hinder, or even threaten our
well-being. Again, peripheral temperature is not simply coupled to
discrete emotions. In fact, when my colleagues and I did a summary
(also considered a "meta-analysis") of the studies of the relationship
between temperature and emotion, we found no convincing evidence
of a causal connection—at least at present. The most likely reason
for this is that psychologists have underestimated the complexity of
the relationship between temperature and emotion. Pure science is
rewarding, but not nearly as remunerative as science applied to the
creation of merchandise. The market for consumer medical diag-
nostics is highly profitable these days, and likely many devices sold
to consumers are quite useful. But I, for one, would not like to see
venture capitalists backing some company that claims the ability to
detect, identify, and distinguish among emotions based on finger
temperature, or, even worse, purports to detect lies based on peripheral temperature measurement. I don't mean this as a dig at corporate

capitalism but rather as an acknowledgment that we psychologists are better at research than we are at giving practical advice and furnishing simple remedies for what ails folks (at least for now).

## We get attached

The Ekman study relied on 12 professional actors and four scientists. In the study, participants were told to contract specific muscles. The idea was that with this *bodily* manipulation, an emotion can be activated in the *mind* without reference to the word or the context. But that context is often vital to our understanding of the emotion. When your partner cheats on you, it will very likely elicit a very different type of anger than when a convenience store cashier shortchanges you. In other words, social context matters.

Just because Ekman and the others take their assertion too far on too little evidence does not mean that their results are without value. Quite the contrary. Physiological changes alone cannot distinguish the complexity of our emotions, but measuring those changes still provides valuable information. The work of Ekman and the others inspired me and my later work on social thermoregulation. Let me explain.

It's important to note that emotions are devices that help us determine our relationship with others—they enable us to understand and predict behaviors. Emotions have a considerable social dimension that serves to navigate our "attachments." On the basis of our current psychological understanding, we can interpret attachment as expectations of our social world. Within this framework of current understanding, the expression of emotion, as through a facial expression, is a *social* action. It communicates to and with the social world. The idea that emotion and its expression should relate to peripheral temperature changes is consistent with *social* thermoregulation. This much is supported by the studies of Ekman and others. That the relationship among emotion, its physical expression, the social impact of that expression, and social thermoregulation is more complex than these studies account for underscores our need for a fuller theory of human social thermoregulation. Function and context cannot be fully defined without understanding their social dimension, which, as we are about to discover, must in turn be understood in terms of our social network.

But we don't want to make the same broad intellectual leap in a single bound that Ekman and his colleagues did. Before we jump further into social thermoregulation and different aspects of our social network, we need to dive into the history and current meaning of attachment. Let's start with a fascinating fact from the world of rats. Rat mothers generate high body temperatures to keep their pups warm and cozy.[3] Although the temperatures are elevated enough to pose a possible threat to the mother's health, it's warmth that the infant rats need. In the absence of their mothers, rat pups exhibit signs of depression or despair. Yet it is not absolutely necessary to reunite the pups with their mother to cheer them up. If the babies are artificially warmed, their observable behavior improves at least somewhat, even in the continued absence of their mother.[4]

The behavior of a rat mother with her infant suggests two things. First, to a human observer, the maternal thermoregulatory behavior of the rat *appears* to be a lot like altruism. Voluntarily or not, she raises her body temperature to dangerous levels to keep her babies warm. Functionally, at least, this is a strikingly social act, one of social thermoregulation. By expending considerable metabolic energy to generate dangerously high body temperatures, the mother, in effect, makes a personally risky "investment" in her progeny. But lest we fall into sentimental anthropomorphism, we must take note of the second thing: the behavior of the rat pups. They need *heat* more than they need their *mother* to generate that heat. Sad when separated from their hot mama, the babies can be cheered, if only somewhat, simply by switching on a warm lightbulb.

Let's explore attachment theory. Its prime assumption is that, provided with any caregiver, infants—including human infants and those of many other species—become "attached"; that is, they become social and capable of forming relationships, beginning with their caregiver (often the mother). The more responsive the caregiver, the more she satisfies the relatively helpless infant's needs and therefore the more secure the infant's attachment to his environment. In infancy, the needs in question are chiefly those on which survival depends: food, protection from predators and other sources of harm, and warmth. Rat mothers risk their lives to provide for their young, especially when it comes to providing warmth. Thermoregulation, then, is an important driver of attachment in infant rats. In a sense, it is even more important than the attachment itself, since artificial sources of warmth will go a long way toward meeting the rat pups' need for heat.

While attachment has profound emotional implications early and throughout life, it is rooted in basic needs essential to survival. It serves an evolutionary end in natural selection. How we attach tells us something about how reliable the sources of warmth were in our early days. The infant rat attached to a competent mother is more likely to survive to maturity and reproduce. Moreover, rodents in general are highly "social" animals; that is, they typically live in groups. Not only is thermoregulation important early in rodent life, but it also seems that keeping warm is among the chief drivers of rodent "society" throughout the life cycle. This becomes especially apparent when cold weather threatens. Research on a Chilean rodent called the *Octodon degus*, or common degu, reveals that each animal reduces its energy consumption by 40 percent yet achieves a higher peripheral temperature when it is (in an experiment) housed with three or five others. To be a solitary degu is to spend much too much energy just trying to keep warm.[5] We can infer that the drive to form "social" groups—a drive we can call attachment—begins for the degu in infancy as attachment to the mother, who provides, among other things necessary for survival, a source of warmth.

The quest for physical warmth associated with attachment in infant rats, infant humans, and infants of other species plays a profound role in attachment behavior later in life. The psychologist Harry Harlow conducted still-famous (and, from the standpoint of animal rights, still ethically controversial) demonstration experiments with infant monkeys in the 1950s. He took infants from their mothers and placed one group with an artificial mother that was a simple, bare-wire model, and placed a second group with a wire "mother" covered with heated terry cloth. Although monkeys "raised" by both kinds of artificial mothers exhibited social deficits as they matured, those who had experienced the bare-wire mother were less socially warm than those whose artificial mother was covered in warmed cloth.[6]

In people, monkeys, and rats, social "warmth" is not a mere figure of speech based on a cognitive analogy with physical warmth. In all of these animals, there is a developmental and physiological connection between the thermoregulation provided in infancy by the proximity of a physically warm mother and the social thermoregulation that drives attachment behaviors later in life.

## We don't huddle; we get more social

As we have seen, humans, like many other animal species, endeavor to attain and maintain temperature-specific homeostasis. This is called thermoneutrality, and it is defined by a difference between core body temperature and peripheral body temperature. In 1847, the German biologist Carl Bergmann described a unique distribution pattern of populations and species. He observed that within a taxonomic clade (group of organisms believed to have evolved from a common ancestor) distributed over a wide area, populations and species of larger animals are found in colder environments and those of smaller animals in warmer environments.

Put it this way: larger animals are generally found farther from the equator than smaller animals, with the largest animals (within the same taxon) found farthest from the equator and the smallest the closest. This pattern has been found to be so consistent since 1847 that it is now called Bergmann's *rule*.[7]

In pretechnological societies and in those situations in which technology is poorly developed or otherwise unavailable, huddling behavior is an efficient way to stay warm. But among modern humans, most huddling is done only in such specific contexts as with your romantic partner or among the members of your sports team. CSI—complex social integration—is a variable often assessed in relationship research. It refers to the creation of diverse social networks. If we look at the literature of relationships and health, it is one of the best predictors of whether or not we remain alive for long (at least in Western, Educated, Industrialized, Rich, Democratic samples).

CSI is, of course, a variable unique to humans. Rats, chimpanzees, and penguins don't volunteer in church groups, join sports teams, or regularly Skype with their friend who lives on another continent. Nevertheless, with respect to thermoregulation, we humans resemble other species in many ways. For one thing, solving the problem of regulating body temperature is for many species, including our own, second in immediacy only to making sure we can breathe. As in other species, our immediate survival requires, first, oxygen and, second, a body temperature regulated within a narrow window of viability. We need water and food, too, but their necessity, though very real, is less immediate. We can survive longer without them, and we can, when necessary, readily store and ration both.

Bergmann himself supplied an explanation for what he observed. He theorized that the ratio of body surface area to body volume is lower in larger animals than in smaller animals, so that larger animals radiate less body heat per unit of mass than smaller ones; therefore, the core body temperature of larger animals remains warmer in cold climates. In warmer climates, however, the organism needs its metabolically generated heat to be dissipated more quickly. Smaller animals, with a higher ratio of body surface to body volume, dissipate excess heat more efficiently than larger animals.

Bergmann's rule illustrates a most basic but profound thermoregulatory adaptation: body size. Along with other large mammals, this thermoregulatory rule certainly applies to humans. Human populations living farthest from the equator and closest to the poles, such as the Aleut, Inuit, and Sami peoples, are generally heavier than people who live nearer to the equator. But both human and nonhuman species do more than rely on this single evolutionary adaptation—size—to address their thermoregulatory problems.

As we saw in penguins, huddling is an effective strategy that depends on the availability of warm bodies—the more the better. Social-network size is therefore key to penguin survival through social thermoregulation. The same has been found to be true for vervet monkeys. Richard McFarland and his colleagues observed that if vervet monkeys have larger social networks, they also have higher core body temperatures when the temperature drops.[8] Similarly, we've seen that a family or group of people can share a bed if they don't have indoor heating or can't afford fuel for fireplaces.

But humans have developed social thermoregulation to far more elaborate levels than other species, for our social evolution has picked up where biological evolution left off. A major difference between human society and "social" groups in other species is that, in our case, *diversity* of social networks is far more significant to thermoregulation than the mere *size* of those networks, and very reliably so. This fact adds to both the interest and the complexity of studying human social thermoregulation, and we do not yet fully understand why network diversity is so important.

What we can observe is that we humans attach to the social world in a wide variety of ways, connecting with a diverse collection of individuals who are capable of addressing various needs—indeed, the entire range of needs and wants that characterize life in an advanced culture, society, and civilization. Nevertheless, the presence of social thermoregulation in humans implies that the original evolutionary impetus for creating diverse social networks was to keep warm in ways that efficiently manage the economy of action. At this level, the social urge of both penguins and people is driven, at root, by the same biological imperative. As colleagues and I have found in recent studies, higher levels of complex social integration elevate the core temperature of people who live in colder climates. This is a fascinating finding, which, like so much else in the study of social thermoregulation, tempts us to commit the logical sin of reverse inference by concluding that people in colder climates are more social than those in warmer climates. Neither society nor psychology works in such neat and simple one-directional ways. Discovering a correlation does not imply cause and effect, but it does imply a relationship—in this case, between social thermoregulation and core body temperature.

The evidence suggests that maintaining core body temperature is a primary evolutionary driver of social thermoregulation and, therefore, of human cultural evolution. Moreover, as we will see later in the chapter, an interesting 2011 experiment suggests the survival-related value of living within a diverse social network. Nevertheless, as of this writing, only a single project exists to demonstrate a relationship between network diversity and human core body temperature. In 2016, Tristen K. Inagaki and her colleagues published a pilot study examining the relationship between physical and social warmth. They concluded that having greater feelings of social connection is positively correlated with a higher core body temperature.[9] It is not clear, however, whether social connections really do protect against physical cold, let alone which specific aspects of social contact provide such protection or, for that matter, whether social networks are better predictors of core body temperature than other known variables. More work is needed.

## The time has come for a credibility revolution

The problem encountered in constructing and interpreting studies that attempt to correlate emotion and temperature with *social* thermoregulation and attachment is, well, complicated. Reality is more complex than so-called emotion theory has hitherto appreciated and accounted for. In fact, reality is so complex that the studies necessary to show that emotions are connected to the experience of emotions *in specific social contexts* have yet to be done!

But let's back up for a moment to how psychologists (including my colleagues and me) arrived at the realization that we needed greater precision in our research. By the time we had conducted some of our first studies on social thermoregulation and had voiced our critique of conceptual metaphor theory, the social sciences, including psychology, were in the throes of what was widely being called a crisis. Many studies published in distinguished peer-reviewed journals reported results that proved impossible to replicate. These included many findings in the conceptual metaphor theory literature cited in Chapter 2, such as Zhong and Liljenquist's 2006 study of the "Macbeth Effect" and a 2012 study by Banerjee, Chatterjee, and Sinha reporting that recalling instances of unethical behavior led participants to see a room as darker and to desire more light-emitting devices (such as a flashlight) compared to participants recalling instances of ethical behavior.[10]

Some social thermoregulation studies have also failed replication. When I attempted to do a follow-up study of the 2008 "coffee-in-the-elevator" Yale study, discussed in Chapter 1, I obtained a comparable result, but it wasn't what scientists see as a close replication. The Yale results therefore required independent verification. I continued more studies myself, basing my predictions on the Yale results. In one study, in which I manipulated warmth and coldness, I confidently predicted that, under warm conditions, people would recall their loved ones. If we "prime" people to be warm, they should think about love, right? It turned out that my prediction was dead wrong. It was in *cold* conditions that people tended to think about loved ones. Moreover, this result proved to be robust. We replicated the effect in large samples in France, not just in the Netherlands, the site of the original studies. I had expected to prove *priming*, the idea that warm conditions would prime people to think "warm" thoughts of loved ones. Instead, I found an effect related to *compensation*. A cold condition produced the cognitive result of social thermoregulation: "warm" thoughts of loved ones.

Perhaps I should not have been so surprised, since Dermot Lynott and colleagues were not able to repeat the Yale results in a replication that closely reproduced the Yale study's procedures.[11] The truth is, we psychologists are often too optimistic about our ability to make predictions and to give practical advice based on them—for instance, heat the room by $x$ to achieve the outcome $y$. The way we conducted studies in the past and the way we analyzed our data simply did not allow us to make very precise predictions for specific individuals in specific situations. But this does not mean that conclusions about the underlying principles and mechanisms are entirely mistaken. Yet if studies cannot be replicated, results cannot be deemed supportive of a hypothesis, an idea, or a notion, let alone a proffered "theory." You can get an idea of just how serious the crisis is when you consider that researchers who published an article in the prestigious British scientific journal *Nature* attempted to replicate 21 social and behavioral science papers published in *Nature* and in *Science*—and were able to successfully replicate only 13.[12]

It was time for a credibility revolution.

It has been a long road, and while we are making progress, we are still not quite there yet. The first changes we (and many other colleagues) introduced was to increase the number of participants from whom we collected data. But even this was not enough, because it allowed for only limited discovery. We needed to think bigger, and that is why, in what we called the Human Penguin Project (HPP)—a major effort to understand the relationship among climate, social integration, and core body temperature—we changed the logic governing how psychologists typically work.[13]

Psychologists usually come up with an idea, often by observation, borrowing from other disciplines or, in some cases, making educated guesses. There is sometimes a struggle to fit research observations into the frame of the idea. We decided to take a different path by

setting out proactively to reduce the incidence of errors that produce results unlikely to be reproduced. Such results are the products
of human error, including unwarranted assumptions of causality;
therefore, to avoid these *human* errors, we would use a *computer
algorithm* to interpret data. Problems with reproducing studies are
often because we mistake "noise" (incidental findings) for "signal"
(a significant, meaningful phenomenon). Called "overfitting," this is
an error that many others and I have identified as playing a significant role in creating the ongoing replication crisis.[14]

Overfitting typically occurs when researchers observe a dataset in
a single context and attempt to apply the granular details to generalize the result with models that are too complex for reality. The fact
is, sometimes things happen due to bad luck. For instance, a participant may not have been interested in answering our questions or
may have come to the lab drunk. Such anomalies, which are part of
life, cannot always be explained by the model we have for our data,
and we should not attempt to force (or overfit) them into that model.

To avoid this problem, we try to apply modern machine-assisted
techniques to generate the most probable model *from* the data rather
than impose our notion of probability *upon* the data. We then also
try to replicate our own results within the dataset. Within our
own research project, we attempt a kind of preemptive replication.
Instead of trying to explain away all our inattentive or drunk participants, we use a technique called supervised machine learning to
conduct an exploratory analysis. This interprets the data without
explicit instruction from us, the human researchers. We do not program the machine to deliver on our prediction. Only after we have
the machine-learning result do we move on to a validation stage.

The HPP was built on a large chunk of prior research—including
our own work—to identify known correlates of core body temperature. Among these were specific social-relationship variables other
researchers had studied, such as nostalgia and attachment to homes.
We also included variables typically thought to affect body temperature, including stress and medication use. Finally, we identified

variables relating to metabolism and network quality, such as daily diet and consumption of sugary beverages, acting on the common hypothesis that relative social isolation leads people to increased levels of sugar intake. We were purposely overinclusive in selecting variables because we wanted to ensure that we were identifying many of the variables that have figured most prominently in the past literature as predictors of core body temperature. We addressed such issues as self-control and attachment as well as alexithymia (the inability to identify and describe emotions in the self), all of which bear directly upon the regulation of stress and, therefore, might relate to body temperature. We factored this array of variables into both our HPP pilot study and the subsequent main study, thereby enhancing the effort at replication.

In the pilot and the main study, participants were asked to complete the survey between 9 and 11 in the morning (local time), to avoid eating or drinking anything warm or cold for 10 minutes before taking the survey, and also to refrain from exercising for an hour before the survey. Participants used an oral thermometer to measure their own temperature before and after taking the survey. They snapped pictures of both the before and after readings, posting them to our online platform.

We began by running the pilot study online with 232 participants. This was a study I ran individually on two crowdsourcing platforms as a "proof of concept." After I had some preliminary yet robust findings, I felt encouraged to recruit friends and new collaborators to collect data in a dozen different countries from 1,523 participants. Psychological experimentation is like political polling: size matters. Many studies fail to replicate results for the same reason that many polls fail to accurately predict election results. Samples are too small and too local, or in some other way insufficiently representative. Our objective was to more accurately identify which variables were most accurate in predicting core body temperature.

To test the social thermoregulation principles we had identified, we needed to test the basics, which required using supervised machine learning to identify both the social and nonsocial factors that bear upon core body temperature. The complexity involved in predicting people's core body temperature lies in the range of variables that must be examined. By allowing the data to drive the analysis in this way, we found that network diversity, defined in terms of the number of high-contact social roles in which one engages, is one key predictor of core body temperature.

Next, we used another procedure to demonstrate that colder climates relate to higher levels of complex social integration and that the higher level of social integration relates, in turn, to higher core body temperature. Despite central-heating technology, people nevertheless rely on social warmth—associated with complex social integration—to counter the physical cold of their environment. We still don't exactly understand how humans do it, but our working hypothesis is that emotions are tied to temperature changes; furthermore, people also help regulate one another's temperature via social emotion-regulation processes. And, as in rats, if you perceive that your interaction partner is emotionally closer to you, you will likely be willing to spend more energy to help thermoregulate him or her.

The first key point to remember is what I have stressed before: don't engage in reverse inference. Just because people farther away from the equator score higher on social diversity does not mean that they are more (or less) social. Humans engage in a variety of social interactions for different reasons, social thermoregulation being one of them. We also share our food and rely on others to feel safe. We cannot infer from our data that one country does better than another when it comes to social interactions. Another key point to bear in mind is that the supervised-machine-learning approach specifies predictors without necessarily implying causality. Assuming predictors are causes is a useful human habit, but a pesky one for scientists. As the saying goes, to the person who has only a hammer, the whole world looks like a nail.

Supervised machine learning gave us a way to explore data meaningfully without presuming any specific causal relationship in testing social thermoregulation principles. Chief among these is the principle that modern human relationships are organized around body-temperature regulation. Sophisticated as our social networks appear today, their thermoregulatory foundation is no different from what it was when it first evolved. This means that social thermoregulation is not only a characteristic acquired from human evolutionary ancestors but is, in terms of its morphology—its form, structure, and operation—unchanged from what it was at an earlier stage of human evolution.

## We predict our safety through attachments

By jumping to explore our diverse social networks, we skipped a big step. Attachment and predicting other people's behaviors is *first* about our very first relationships with our caregivers, and *then* about our most intimate relationships. In the HPP, we found and replicated the idea that protection against the cold via more diverse social networks applies only to those in a romantic relationship. Although we are not yet sure why, we do have a considerable volume of data to provide us with some pretty good ideas.

Thanks to the work of John Bowlby and Mary Ainsworth, along with the experiments of Harry Harlow, it is now widely believed that attachment styles are shaped beyond infancy and that we apply our attachment styles to other relationships in adulthood. Harlow's experiments with monkeys "raised" by artificial mothers provide a simple illustration of this principle. Infant monkeys who had only cold, bare-wire artificial mothers to cling to matured as socially cold adults. Those who had artificial mothers made of wire wrapped in warmed terry cloth matured into relatively warmer adults, in social terms, though they still suffered a social deficit compared to monkeys raised by a live mother. In humans, attachment styles are necessarily more complex because of the diversity of relationships in which we engage, but they form a basis or model for relationships far beyond that between infant and caregiver.

As useful as the concepts of attachment, attachment styles, and the classification of these styles have been, especially to clinicians and therapists, there is still work to be done toward understanding the biological mechanisms of attachment. A fuller understanding of biological origins is crucial to devising effective clinical and therapeutic interventions. Rat research, as we have seen, suggests that the evolutionary driver of rodent group behavior is the need to be safe from predators and to defend against cold. In species whose young require parental care and feeding to survive—true of rats and humans alike—defending against predators and the cold is especially urgent. Natural selection has favored attachment behaviors because they promote survival to a reproductive age.

Related to protection against predation and cold, attachment is all about distributing risk among social groups and, you guessed it, thermoregulation. Arguably, in modern humans, whose cultural evolution has manifestly extended the adaptive benefits of biological evolution, the biological drivers of attachment and complex social integration should loom with less importance in adulthood. This, however, is not the case. The combination of early dependency and the role of infant dependency throughout evolution, as we saw in the HPP, means that human adult social life is *still* organized around the two primal objectives of risk distribution and body-temperature regulation. Remarkably, this remains true even though cultural evolution has spawned a variety of economic, political, moral, and technological defenses against a wide array of environmental threats, including predation and cold.

As we noted at the beginning of the chapter, studies by Ekman and his colleagues (along with others) suggest that the act of making an emotional face—sad, angry, and so on—in response to an experimenter's request is associated with measurable effects on the autonomic nervous system (ANS), including changes in peripheral temperature. Such ANS effects, however, are also related to social thermoregulation and not exclusively to simply defined discrete emotional states such as anger, fear, and sadness. Emotions and their physical expression occur not just within an individual but within the social context of a relationship. Thus, emotions and their expression are related to attachment.

The psychological literature on "attachment theory" is extensive and continues to evolve. Although I think that the study of attachment is probably one of the most advanced areas in most of psychology, it remains presumptuous to speak of a "theory" in psychology. The humbling truth is that our discipline itself is not yet sufficiently mature to serve as a context for theories. A theory, in the true sense of the word, must have formal predictions, and this is something most of our subdisciplines do not yet have. At most, we have ideas— call them principles—built on correlations and a collection of causal relationships derived from experimental studies. This said, the *principle* of attachment is highly compelling. We must, however, recognize that as a "principle," attachment is a subset of "interpersonal relationships," not a synonym for that phrase. In a 2005 critical review of studies on attachment, psychologists Everett Waters, David Corcoran, and Meltem Anafarta pointed out that attachment "theory" is not intended to be a general theory of relationships but rather an explanation of how humans respond in relationships under such stresses as separation from loved ones, injury, or perceived threat.[15]

I would rephrase this. Attachment is first and last about adapting to your environment. Life-history theory is a theory of biological evolution that originated in the 1950s to explain various aspects of an organism's anatomy and behavior by looking at how the organism's life history—especially its reproductive development and behavior, postreproductive behavior, and life span—has been impacted by natural selection. With respect to attachment, life-history theory teaches us that the harsher the environment in which an organism develops, the less secure the organism's attachment. This is because, in terms of natural selection, survival in a harsh environment depends on a high degree of vigilance, and vigilance is an approach or a posture incompatible with the complacency that secure attachment implies.

Recall the experiment from Chapter 1 in which we found that children in a warm room donated more stickers or balloons to a (fictional) "kid next door" than did those in a cold room and that, furthermore, this effect was mostly seen in children who felt securely attached. As the replication crisis hit, however, we came to realize that our sample sizes were too small to warrant meaningful conclusions. Moreover, in the meantime, some studies had come out that appeared to show no effect between attachment and social thermoregulation.

So we decided to embark on a new project consisting of three studies based on a rationale quite similar to the HPP. We presented people with 36 statements about their feelings of attachment in close relationships. Typical items were "My romantic partner makes me doubt myself," "I feel comfortable sharing my private thoughts and feelings with my partner," and "I find it easy to depend on romantic partners." Participants responded to the statements on a scale from 1 (completely disagree) to 7 (completely agree). Our participants then touched a warm—or cold—cup, and we asked them to name five people who spontaneously came to mind. After they did this, we asked them how close they felt to each individual they named. Based on earlier studies conducted at Yale in 2008 (in which participants were asked to rate people in terms of social warmth or coldness after they had held either warm or cold cups), we predicted that holding a warm cup would elicit thoughts about people to whom our study participants felt closer.

We were very wrong. People who had held cold cups thought mostly about loved ones. Moreover, we consistently replicated this effect. Participants in the cold condition consistently thought of loved ones.

We took the analysis a step further. As in the HPP, we wanted to explore whether and how attachment to others affected participants' responses to temperature. Previous attempts to do this had yielded inconsistent effects. By presenting our 36 statements followed by contact with a warm or cold cup, a request to think of five people, and our questions about how close participants felt to each of the five, we were able to detect a *consistent* pattern. Those participants who believed they could depend on their partner and felt comfortable sharing with him or her their innermost thoughts tended to think more of loved ones when cold. The reverse occurred for those who did not feel comfortable depending on romantic partners. An intriguing wrinkle in this study is that I had in the meantime changed jobs and moved from the Netherlands to France. This allowed me to test the same idea with a very different group of participants: French as opposed to Dutch students. The third study was conducted by my new French collaborators, Lison Neyroud and Rémi Courset, among students in Grenoble living closer to the equator. Although we found the same effects in this group of students, they were—in line with the HPP—somewhat less sensitive to the temperature manipulation.[16]

What is true in nonhuman species is readily extrapolated to attachment in humans. A person who grows up in a harsh or threatening environment tends to become a vigilant, perhaps hypervigilant, individual with insecure attachment. Most of us have an understandable inclination to assume that secure attachment is emotionally "healthy" and therefore "good," whereas insecure attachment is inherently "unhealthy" and thus "bad."

But let us consider the ostrich. To find food and eat it, ostriches must keep their heads down close to the ground. When an ostrich does this, it cannot also be on the lookout for predators. In other words, to forage is to risk your life—especially if you happen to have the superlong neck of an ostrich. To avoid the risk, however, is to starve. Ostriches mitigate the hazards involved in eating by distributing the risk to a group. While some ostriches are foraging and eating, others in the group have their heads up and are in a position to respond to predators.[17] Their response serves to alert the entire group. In ostrich "society," risk is distributed. While some of the birds in the group complacently peck close to the ground, the more anxiously attached ostriches remain vigilant because they predict that they cannot consistently rely on others. In this way, the anxious attachment of some ostriches can be helpful to the entire group. In fact, it is a matter of life and death.

An experiment conducted in 2011 with humans, not ostriches, suggests the potential life-and-death value of being a member of a large—and diverse—group.[18] Researchers "unobtrusively observed" a total of 46 groups who were each put in a lab room that gradually filled with smoke, apparently coming from a malfunctioning computer. If even one person in a group was characterized by attachment insecurity—was anxiously attached—the entire group was more likely to leave the room than if they all felt secure. That is, as the researchers put it, "attachment anxiety was associated with quicker detection of the danger and . . . speedier escape responses to the danger once it was detected." The safety of the group was enhanced when the group was sufficiently diverse to include an attachment-avoidant individual. Indeed, when no one in a group demonstrated anxiety, the group was less likely even to notice the smoke. Had the staged emergency been real, a happily homogeneous group of securely attached individuals might well have died, complacently attached to their lethal environment, because they made the wrong predictions. As Benjamin Franklin said, "If everyone's thinking alike, then no one is thinking."

## Attachment is about managing metabolic resources

Thus far we have alluded to prediction and energy conservation, but we have yet to get into the details. In the previous chapter, I mentioned work by my friend Jim Coan. In 2014, I visited him at the University of Virginia, in Charlottesville. It was a visit that profoundly changed my perspective on the field of social thermoregulation.

I picked up on an epiphany Jim had had years earlier. He was doing studies on what he calls the "handholding effect." He put people, mostly women, in an fMRI scanner and informed them that they would periodically receive a mild electric shock to the ankle. As Jim knew it would, this made the participants apprehensive. He wanted to measure stress-related brain activity with the fMRI scanner under different conditions: with a hand-holder who was a partner, with a hand-holder who was a stranger, and with no hand-holder at all. Before beginning the study, he sensibly predicted that when the partner was holding the participant's hand, the participant would become better at regulating her emotions. He predicted that the fMRI would show the brain "lighting up" more in areas typically related to stress, such as the prefrontal cortex, with a hand-holder present than without. After all, the dominant model at that time was that people believed they could rely on others as resources for social support.

Jim Coan conducted the study, and what he found was exactly the opposite of what he had expected. When a partner was present as the hand-holder, those brain areas related to stress were lighting up *less* (and this was the case only a bit less when the hand-holder was a stranger).

Jim had some explaining to do, especially because he kept finding the same results. At that time, Dennis Proffitt, a cognitive scientist at the University of Virginia, offered an explanation as simple as it was elegant. The participants in Jim's studies, he said, were behaving like ostriches. They were not doing *more* emotional work but less, because they were distributing (off-loading) the stress. With this realization, Jim's social baseline principle was born: our baseline expectation is that we are with other people. If we are alone, our baseline expectation is not met, and we feel threatened.[19]

In essence, this is what it means to be *social*. It is a profound and stunningly simple insight. Proffitt's inference was based on many studies he had conducted on the idea of the economy of action. Proffitt's work on the economy of action was, as you'll recall from earlier discussion, based mostly on studies involving the estimation of distance and hill slant. In judging hills, people tend to overestimate slant when describing it verbally, expressing the incline in degrees, or visually. Proffitt (and the many colleagues with whom he conducted this research) reasoned that our estimates are in service of our engaged actions. If we overestimate, it is our brain and body's way of telling us not to climb that hill.

The Israeli researcher Tsachi Ein-Dor subsequently conducted studies with Jim Coan on a closely related topic. To assess the consequences of being avoidant in one's attachments—that is, in how much people feel they can rely on others—he measured fasting glucose in their blood. He found that those who are more avoidant tend to have higher fasting glucose levels. The researchers concluded that this indicated that avoidant people needed to mobilize greater metabolic resources. After all, depending only on oneself takes more energy than distributing the work socially.[20]

## Co-thermoregulation is also about attachment

When it comes to attachment, most of what psychologists know about and measure relates to distribution of risk. For example: Do I feel safe and secure with my partner and can I rely on him or her when the going gets tough? But what about individual differences in our desire to socially thermoregulate? Social psychologists, my beloved colleagues, have always been obsessed with the power of the situation, often to the detriment of understanding individual differences. But one of the wonders we have to celebrate about life is that people differ in many ways, including in their desire and ability to socially thermoregulate. Bigger animals tend to preserve body heat better and, conversely, have a more difficult time when the ambient temperature gets really hot.

We likely also differ in the degree to which we had reliable thermoregulatory care in infancy. So, from the same dataset of 1,523 participants in 12 countries used in the HPP, my colleagues and I set out to develop a measure of individual differences. This measure, called the Social Thermoregulation and Risk Avoidance Questionnaire, or STRAQ-1, also drew extensively on prior research.[21] The STRAQ-1 project was specifically designed to address a gap in our understanding of attachment both in infancy and in adulthood, and to better understand the relationship between environment and personality. Recall that attachment—the bond formed between an infant and caregiver that drives much subsequent social, emotional, and cognitive development—is explained by an assumption that infants, who are relatively helpless, rely on others to survive. An important aspect of this assumption, which is not always sufficiently emphasized, is that survival depends on the infant's ability to successfully cope with environmental demands. Our infant bodies necessitate reliance on others; warming up by ourselves, because of our modest body size, is impossible.

STRAQ-1 explored the idea that the quality of interpersonal attachment as manifested in adulthood is strongly related to whether an individual, in infancy, is able to successfully distribute temperature regulation to his caregiver or caregivers. If the caregivers reliably meet the infant's need to regulate temperature, we can expect that the maturing infant will develop secure attachments throughout childhood and into adult life. The questions in the STRAQ-1 were designed to assess the participant's tendency to distribute body-temperature regulation.

Proceeding from our conjecture that the social regulation of temperature and stress is formative in attachment styles, we analyzed the STRAQ-1 data and identified as relevant 23 (out of 57) items in four subscales: social thermoregulation, high-temperature sensitivity, solitary thermoregulation, and risk avoidance. The participants who score higher on questions relating to the desire to socially thermoregulate—cuddling with someone when cold—tend to be in better health, and, consistent with the idea that reliance on others should relate to having confidence in them, they are less avoidant in their attachments to loved ones. Avoidance in attachment was measured with the Experiences in Close Relationships questionnaire and included such statements as "I prefer not to show a partner how I feel deep down," "I don't feel comfortable opening up to romantic partners," and the reverse-scored statement, "I talk things over with my partner." People who agree more with the first two statements tend to "predict" that others "will not be there for them" when they need them emotionally.

Unsurprising to us, the people who tend to rely on others less for social thermoregulation tend to feel less comfortable relying on and opening up to their partners. These people also tend to externalize their emotions more, essentially reporting that they recognize their own emotions less.

Here are some of the STRAQ-1 items relevant to social thermoregulation, so that you may think about it for yourself:

> *I usually have more physical contact with others than most people.*
> *When people are close to me, I like to be really close to them.*
> *When I feel cold I seek someone to cuddle with.*
> *I like to warm up my hands or feet by touching someone who I am close to.*
> *I prefer to warm up with someone rather than with something.*

Those who score higher on high-temperature sensitivity questions include older people, people who are more stressed, people with smaller social networks, and people who are poorer at recognizing their own feelings. High scorers on high-temperature sensitivity tend to be more anxious in their attachment. These items are typically drawn from the same scale as the avoidance items and include such statements as "I'm afraid that I will lose my partner's love," "I worry a lot about my relationships," and "My desire to be very close sometimes scares people away." The attachment systems of people who score higher on these items are typically overactivated. *They* are the ones who are vigilant in the room with a smoking computer. People who are more anxious tend to "predict" that people will probably be there for them but not consistently. Some of the key survey statements relating to high-temperature sensitivity include:

*I find warm days pleasant.*
*I find hot days pleasant.*
*I don't like when it's too hot.*
*When I feel warm I do not want to do anything.*
*I can't focus when it is too hot.*
*I prefer to relax in a cold place.*
*I am sensitive to heat.*

On the solitary-thermoregulation statements, taller people score lower. The higher-scoring people include those who report higher levels of stress, tend to feel more nostalgic, externalize their emotions less (recognize their emotions better), and are generally also more anxious in their attachment. STRAQ-1 items relating to solitary thermoregulation are:

*I am not sensitive to coldness.*
*When it is cold, I more quickly turn up the heater than others.*
*When it is cold, I wear more clothing than others.*
*I can't focus when it is too cold.*
*When I feel cold I don't turn on the heater.*
*When I am troubled I like to take a long warm shower to clear up my thoughts.*
*A warm beverage always helps me relax when I am down.*
*If I am feeling distressed, I seek a warm place to calm down.*

The seemingly most robust finding was that social thermoregulation has a negative correlation with attachment avoidance. In other words, people who score higher in social thermoregulation are less avoidant in their attachment. It is important to recognize that the effects we have described are correlations. We can infer that they are directional—that a low level of social thermoregulation can cause you to avoid attachment. But we have no proof of this cause and effect, which means that more studies need to be done, particularly ones that follow people for an extended period of time.

We can also dare to infer that attachment, with respect to social thermoregulation, helps the brain function as a prediction machine, forecasting what we called "social weather" in Chapter 1. Recall the studies revealing that, if researchers make us feel lonely, we estimate ambient temperature as being lower. Attachment guides us in the long-term planning and management of our complex social integration (CSI). If we are securely attached, we invest in the diversity of our social network in much the same way that Warren Buffett invests the Berkshire Hathaway portfolio—for the long run. We are confident in the reliability of others to keep us warm and help us avoid danger. We freely distribute thermoregulation (emotionally investing in people we predict to be "warm") as well as risk (we predict that our diverse social network includes people who have our back). Such social investments are made for the long run. If we are securely attached, we don't look to make a quick killing but to secure long-term dividends. In contrast, if we are anxiously attached, we invest in CSI in the manner of Jordan Belfort, the "Wolf of Wall Street." Anxious, our senses hyperactivated, doubting the reliability of others and therefore unable to predict their "warmth," we invest for the short term. We look for an instant return on our social investments. Sometimes we place a good bet and do very, very well, but we never take steps to build up our portfolio for the long term.

Altogether, there is decent evidence for the idea that other people are crucial in supporting our ability to thermoregulate. Of course, before our culture created technologies of efficient heating, people huddled for warmth. The need to up-regulate body temperature drove ancient human relationships. The HPP results are consistent with the idea that, in modern human relationships, we still predict other people's trustworthiness and reliability by gauging whether they are socially "cold" or "warm." Like ancient huddling, modern diverse-social-network creation raises core body temperature.

Attachment was a biological evolutionary adaptation that was extended through ancient cultural evolution into modern attachment in a world of complex social relationship-building. This continuity may be compared to the continuity from attachment in infancy, since it is driven by the urgent physical need for warmth from closeness to a caregiver, which then continues into adulthood. Adults *still* seek warmth, not from physical proximity to a caregiver but through a romantic partner and integration into complex social networks. Evidence, then, is robust for continuity between attachment in ancient cultural times and that in modern times, as well as between infancy and adulthood. In this sense, the gap in our understanding of attachment has been narrowed—except that we still do not understand the mechanism by which CSI increases body temperature.

That doesn't mean we cannot formulate a hypothesis. *Coregulation* is a term psychologists use to describe an individual's continuous action or behavior that is modified by a partner's continuously changing actions or behavior. (We will get back to coregulation at the end of this book.) Although the mechanism by which CSI increases body temperature is not yet understood, we do have reason to believe that it is coregulatory or, more specifically, co-thermoregulatory. A 1969 study by V. Vuorenkoski and colleagues suggests that a mother's peripheral temperature increases when her infant is in distress.[22]

I am involved in a 2014 (to date unpublished) study in which we found that an individual's peripheral temperature increases when they see that their partner is sad.[23] Arguably, the rises in temperature (reported in 1969 and in 2014) are instances of coregulation. We are still in the process of testing whether the latter effects hold up in later replications. Moreover, the psychological literature on relationships exhibits more instances suggesting that people physiologically coregulate to achieve homeostasis, which presumably includes temperature homeostasis. Regulation—and possibly coregulation—of temperature is implicated in attachment and therefore may be the basis on which people form predictions about others and thereby build social-network diversity.

## Predicting the social weather

Psychologists like vervet monkeys not because they are exceptionally cute (which they are) but because they share with humans a number of less attractive characteristics, including a tendency toward hypertension, anxiety, and both social and dependent alcohol use. Like people, they also exhibit a drive toward creating large social networks, which they use, apparently, as a Warren Buffett–style long-term economic investment in energy. We noted in Chapter 3 that vervet social behavior such as grooming significantly insulates the animals against the cold. Indeed, a vervet's fur is such that even grooming (by back-combing) a pelt that has been removed from a dead vervet increases the temperature beneath the pelt.

Vervets, like penguins, "know" the importance of forming large social networks. In effect, early in life—presumably, like human infants clinging to the caring warmth of their mothers—some part of the vervet brain becomes a "prediction machine" by which the animal comes to know that it can rely on others to keep it warm through a combination of proximity and grooming. Researchers have measured the body temperatures of vervets in large social networks as well as the temperatures of more solitary vervet monkeys. The highly attached vervet within a large network has a lower body temperature than the lonely vervet. Attached to vervet society, the animal predicts that, when needed, others will be there to warm it. Although we do not have precise knowledge of the mechanisms involved, we can infer that attachment in infancy, developed and confirmed by social experiences in maturing life, "turns down" the well-connected vervet's thermostat, lowering its normal body temperature and thereby conserving metabolic energy. Vervet "society" exists—at least in part—to facilitate the economy of action, something to which each individual monkey contributes and from which each benefits.

What drives vervet "society" drives human society as well—with an important difference. In vervets (and in penguins), the size of the network provides highly economical protection from the cold. In humans, diversity—far more than the size of the social network—provides this protection, not only giving members the emotional feeling of warmth but actually correlating with measurably higher oral temperatures.

As with much else in psychological science, further study is required to understand the mechanism by which the quality of a social network is correlated with up-regulating body temperature. Meanwhile, as the state of understanding currently exists, we believe that the correlation between a diverse social network and measurably higher oral temperature is a function of coregulation. In a 2012 article, Emily A. Butler and Ashley K. Randall offered a more operational definition of coregulation as "a bidirectional linkage of oscillating emotional channels (subjective experience, expressive behavior, and autonomic physiology) between partners, which contributes to emotional and physiological stability for both partners in a close relationship."[24]

How do these oscillating emotional channels actually come to be linked? We don't know—yet—but we have experimental reason to believe that social thermoregulation holds the key to explaining the connections between attachment and coregulation in the context of both biological evolution and cultural evolution. Operating at the level of the smallest possible social group, coregulation hints at something we don't yet have: an understanding of the biological mechanisms on which animal groups and human societies are built. This is where we begin the next chapter.

## CHAPTER 6

# Not by Hypothalamus Alone

## *How Culture Transforms Social Thermoregulation*

How did we humans go from penguin-style huddling to interacting in very complicated and diverse social networks that help protect us from the cold and have nothing to do with keeping each other physically warm? That is the question I ask in this chapter. Or, to put it another way: How did we get from penguins and rats to people, intensely social beings who built an increasingly diverse network called civilization, extending it even into cyberspace?

### Gezelligheid: Warmth without the penguin huddling

When I lived in the beautiful city of Haarlem in the Netherlands, I often worked in coffee shops. I am, of course, far from alone in this social habit. J. K. Rowling wrote much of her first Harry Potter book in the Elephant House, a coffee and tea establishment in Edinburgh. The global coffee chain Starbucks has over 28,000 locations thanks in large part to customers who enjoy working there. Recall the warm-cup studies that my collaborators and I did in Amsterdam, Utrecht, Tilburg, and Grenoble. The commercial genius of Starbucks is that it serves those hot beverages in settings many people find convivial. Starbucks founder Howard Schultz says that the idea for Starbucks came to him in Italy, a country often credited with having "invented" café culture. Admirers say that Starbucks combines the unique social warmth of the traditional Italian café with the mass-market convenience of McDonald's.

I must admit that Starbucks has never worked its magic on me. It might borrow one element from an Italian coffeehouse and another from McDonald's, but no Starbucks could ever capture what my favorite place in Haarlem has. Its name is Native, and it is what we Dutch call *gezellig*. That translates very loosely as "cozy," and its noun form, *gezelligheid*, translates as "coziness." But a lot is lost in translation. The Dutch word specifically conveys a kind of social warmth. Let me put it this way: when I am nostalgic for gezelligheid, I reminisce about Christmastime, when it's cold outside and we are gathered with the family around the fireplace in a house fragrant with good things to eat and drink. While *gezellig* remains largely confined to the Netherlands, the Danish *hygge* has reached the U.S. public, and that word comes closer to *gezellig* than *cozy* does, in my opinion.

Native was a great place to work. I felt warm, positive, and energized there. When I moved from Haarlem to teach at Université Grenoble Alpes, I found the people hospitable, the food amazing, and the natural landscape incredible. The only thing missing was my coffee shop—so aptly named Native. I realized that, for me, the feeling created by being in my familiar hangout was not just from the coffee. Nor was the feeling anything nearly as essential or intense as being with your loved one in the intimacy of a close relationship. It was somewhere between these two states: a place that offered gezelligheid as well as a connection to the local community. Even as the warmth of the coffee itself bridged the domains of physical and social warmth, a place like Native was a portal offering connection with the Haarlem community.

Fortunately, before too long, my better half discovered a place in Grenoble called Brûlerie des Alpes, where I now go to work every Wednesday. I've actually been doing most of my research for this book there. The coffee is amazing, and the place is run by a family, one of whom, Julien, has offered me wonderful restaurant recommendations. He's revealed that the best "cave à vin" in France is in Charavines, just 20 minutes from my house. He and I have discussed matters of the French economy, including the *gilets jaunes*, the members of the "Yellow Vest" movement agitating for social justice in France, and I—who was about to get married at the time—learned that *enterrement de vie de garçon/jeune fille* (translated as "burial of the life of the young man/woman") is the French phrase for a bachelor or bachelorette party.

Where else could I get such diversity of social information? Brûlerie des Alpes, like Native, is built around serving, consuming, and sharing coffee, a physically warm beverage that conveys social warmth. Both places are also very socially connected and connecting. These places enhance, amplify, and extend the social "content" of their coffee in ways that, for me, Starbucks simply cannot begin to achieve. Both places seem to join the physical, biological, and evolutionary roots of thermoregulation with its cultural stems, leaves, and flowers. They are a cultural expression of the social warmth we crave and, if we are fortunate, often experience. It is the warmth of social-network diversity, without our having to huddle like penguins.

## Evolution hits humans, too

Natural selection has resulted in a variety of interesting thermoregulatory adaptations, and I'd even go so far as to say that the need to thermoregulate has been among the major drivers that have made humans the species it is today. In Chapter 5, I mentioned Bergmann's rule, which holds that larger animals are generally found farther from the equator than smaller animals, with the largest animals (within the same taxon) found farthest from the equator and the smallest found the closest. Bergmann's own explanation for this rule was that the ratio of body surface area to body volume is lower in larger animals than in smaller ones, so that larger animals radiate less body heat per unit of mass than smaller animals. This means that the core body temperature of larger animals remains warmer in cold climates. Conversely, in warmer climates, there is an adaptive advantage for an animal to dissipate metabolically generated heat more efficiently. Warmer conditions therefore favor the survival of smaller animals, whose higher ratio of body surface to body volume makes for a better radiator.

Lending further support to Bergmann's rule is another "ecogeographical" rule, this one formulated in 1877 by the U.S. zoologist Joel Asaph Allen. Allen's rule states that homeothermic animals have shorter limbs and other body appendages (like mouthparts or sexual organs) in colder climates than in warmer climates.[1] The German zoologist Richard Hesse proposed another rule to extend Bergmann's rule. Hesse observed in 1937 that animals of a given species living in colder climates have a larger heart in relation to body weight than animals of a closely related species inhabiting warmer climates. This is known as Hesse's rule or the "heart-weight rule."

All three evolutionary, adaptive rules also apply to humans. For instance, they may account for why the East African Maasai have notably slender bodies and why the legs of the average Belgian are not nearly as well suited to high-jumping as, say, those of the typical member of a Native American tribe indigenous to Arizona. Of course, the ratio of the body's surface area to volume and the size of the heart in relation to body weight are not the only adaptations humans manifest in the service of thermoregulation. We vary in how we physiologically manage heat and cold, with men and women differing in how they deal with temperature. As it turns out, the major factor that explains why women feel cold more readily than men is height. Men are usually taller and thus conserve body heat more than women do. Irrespective of gender, individuals also differ in the ways their blood vessels constrict in response to cold conditions, in their basal metabolic rates, and in the amount of brown adipose tissue (BAT) they store, among other ways.

## Baby fat, a big brain, and a narrow pelvis

Among the universal characteristics evident in human evolution is the anatomical feature popularly called "baby fat," which has much to do with BAT. Subjectively, baby fat is one of the features of infants that make them cute and attractive, perhaps enhancing the impulse many of us have to embrace and cuddle babies. This cuteness may therefore be a quality of adaptive importance. Far more objectively measurable, however, is the fact that BAT makes up some 5 percent of a newborn's body mass and is especially prevalent on the back, along the upper half of the spine, and up toward the shoulders.[2] The most obvious inference is that the presence of body fat helps to avoid hypothermia both because it acts as insulation and because it stores energy. There is an urgent reason why human babies who are born prematurely almost always require extra warmth and, in a hospital setting, are typically housed in incubators. Hypothermia is a leading cause of death in premature infants. "Preemies" are not only smaller but thinner than babies born at or near term.

Indeed, healthy full-term human infants are quite big and fat, compared to infants of most other species. This distinction is even found far back in evolution, in fossil evidence, and these adaptations were necessary for the earliest hominins (ancestors of modern humans), who evolved in Eastern Africa. "Lucy"—3.2-million-year-old hominin fossils representing about 40 percent of the skeletal

remains of a species called *Australopithecus afarensis*—was discovered in 1974 near the Awash River in Ethiopia. In March 2015,
researchers discovered in the same area a 2.8-million-year-old fossil
of what they believe is the lower jawbone of an individual belonging
to *Homo*, the genus to which modern humans, *Homo sapiens*, also
belong. The desert climate is hot during the day, and *Homo* lost a
lot of body hair in an adaptation that enabled survival in the heat.
Dry desert climates show wide variation between day and night temperatures, however, and the loss of body hair presented a challenge
to survival in the cold desert nights. Baby fat and relatively large
neonatal size were heat-saving evolutionary adaptations that helped
to compensate for the loss of hair.[3]

As with the large nose and nasal cavity in the much more recent
ancestor, *Homo neanderthalensis*, baby fat and body size were not
sufficient adaptations to provide a complete thermoregulatory survival solution. Evolution also selected for bigger brains (especially
the cortex). Increased cognitive capacity gave members of the genus
*Homo* a greater ability to plan action for regulating temperature in
advance: finding a different shelter, gathering fuel for fires, hunt-
ing to provide a food reserve or hides for clothing, and so on. More
brain power enhanced the ability to create a metabolic budget to
better manage heat and cold. Executing plans requires a good deal
of activity, and human beings evolved to be far more active than
other species.[4] Bigger brains enabled greater activity, which in turn
promoted selection for bigger, more powerful brains.

Of course, thermoregulation was not the sole adaptive advantage
driving selection for bigger brains any more than that it was the only
factor behind the evolution of the endothermic versus ectothermic
condition. We may speculate that thermoregulation was the pre-
cipitating factor in the evolution toward endothermy, but there is
no definitive evidence of this. The current belief is that the movement toward endothermy was most directly linked to the development of high activity sustained by an aerobic metabolism.[5] But

being highly active is also a plausible response to the pressures of thermoregulation—among other things.

As usual, science has not yet produced sufficient data to reach definitive conclusions other than that evolution, in both its genetic and cultural dimensions, is not only enormously complex but also a moving target, even if it is moving slowly. There was a time, for instance, when many researchers were prepared to conclude that bipedalism, one of the distinguishing characteristics of human beings and their relatively recent ancestors, evolved directly in response to thermoregulatory needs. In a 1991 paper, Peter E. Wheeler argued that "the thermoregulatory advantages conferred by bipedalism to a large-brained primate on the African savannah could have been significant factors contributing to the adoption of this unusual mode of terrestrial locomotion." He cited "a dramatic reduction in direct solar radiation exposure" and other advantages resulting "from the higher distribution of the body surfaces." Because wind speeds are greater and air temperatures lower farther from the ground, upright bipeds enjoy a higher rate of heat dissipation by convection. Above the level of surface vegetation, increased airflow and lower relative humidity promote the rapid evaporation of sweat, which also increases the efficiency of heat dissipation.

Today, most researchers discount so direct a causal relation between thermoregulation and bipedal evolution, but they do point out that selection for bipedalism had other consequences linked to thermoregulation, particularly the loss of body hair, which certainly made dissipation of heat more efficient.[6] Bipedal locomotion is also associated with the evolution of a narrower pelvis, which was advantageous for walking on two legs more efficiently. The greater efficiency of movement is linked to greater activity, including that related to thermoregulation and planning to stay warm or cool off in anticipation of seasonal or other changes in habitat. Interestingly, skeletal fossils suggest that Neanderthal neonates had bodies and heads quite similar in size to the newborns of modern humans.

This seems to imply that giving birth was probably no easier for Neanderthal mothers than it is for the mothers among modern humans. Some researchers conclude that this anatomical circumstance, a product of genetic evolution, created the need for midwives or their equivalent—a rudimentary cultural/social innovation. However, the pelvic form started changing only about 200,000 years ago when *Homo* evolved in Africa and the Middle East. Studies of fossil hominins reveal that those from warmer climates, including the earliest *Homo* examples from Africa, had smaller pelvic breadths than hominins from colder climates. This is an example of conflicting selective evolutionary pressures. While greater pelvic width is beneficial to accommodate large head size, a narrower pelvis makes for greater efficiency when walking upright on two legs and is beneficial to thermoregulation in hot climates.

In climates closer to the equator, temperatures could be so high that a narrower birth canal was necessary to increase heat dissipation, which was not an imperative for the Neanderthals, who lived much farther from the equator. Pelvic evolution in hominins living closer to the equator represents a morphological compromise. The capacity of the birth canal increased along the front-to-back dimension but narrowed in width. In this way, passage of a larger-headed baby could be accommodated, but the infant had to make a difficult turn through the birth canal. Although the Neanderthals of colder Pleistocene Europe appeared on the scene later than their ancestors in the tropics and had wider pelvises, their anatomy nevertheless reflected the much earlier evolutionary compromise in the shape of the birth canal. Neanderthal midwives may well have been a necessity to aid a newborn's passage into the world.[7]

Evolutionary narrowing of the pelvis may also be linked to the necessity of brain development after birth. Human infants are profoundly incapable of caring for themselves, requiring intensive care, feeding, and other attention from the mother and others. This is not only a matter of immediate survival, but it is also a means by which the immature brain is preparing to participate in culture and the environment outside of the womb. Thus, genetic evolution quite literally delivers the newborn into a world in which cultural evolution will play a hugely formative role.

## Thermoregulation lays the foundation for building culture

In large mammals such as humans, homeothermy—maintenance of a stable, optimal body temperature—is a sign of good health. Homeothermy appears in animals that are adequately nourished and hydrated and energetically uncompromised. These conditions, in turn, suggest an animal that responds optimally to environmental challenges. Conversely, heterothermy is linked to poorer health associated with inadequate nourishment and hydration, implying that an individual incapable of maintaining "normal" body temperature is suboptimally adapted to the challenges of its environment. It is likely that such individuals would have lower reproductive success than those exhibiting more adequate homeothermy. Thus, heterothermic individuals would tend to be culled out of the gene pool, their nonadaptive genetic characteristics eliminated over evolutionary time.

We have reason to believe that our close evolutionary cousins, the Neanderthals, had some adaptations specifically suited to the cold environment of Pleistocene Europe. In a 1952 paper, the U.S. anthropologist Francis Clark Howell proposed that aspects of the Neanderthal face were explicitly thermoregulatory adaptations, in particular the large nose and nasal cavity, which served to warm the air that was breathed in. Others subsequently proposed this as the "nasal radiator hypothesis."[8] By warming incoming air, they said, the projecting nose and large nasal cavity helped to protect the brain against suboptimally cold temperatures. Another Neanderthal adaptation was increased activity, which, among other things, helped to generate more body heat. In this case, the large nose and nasal cavity may have helped to cool the face during intense exertion by minimizing respiratory moisture loss in cold, dry environmental conditions. The notably large Neanderthal chest likely allowed individuals to generate heat but was not conducive to preserving heat. The large chest was part of a stocky body, an anatomical adaptation that offered some thermoregulatory benefits as well as strength but, like the large nose and nasal cavity, was not in itself adequate to maintaining viable body temperature in cold conditions without increased activity.

Recall that our exceptionally high level of activity is one trait that differentiates us from other animal species. If the Neanderthals had not found a combination of anatomical adaptations that somehow perfectly suited them to life in cold Pleistocene Europe, especially in the winter, they might never have developed technologies such as clothing, or techniques for creating controlled fires, to extend biological evolution into cultural evolution. According to a 2006 paper by S. E. Churchill, the "lower critical temperature (the temperature at which an unclothed human must increase internal heat production to maintain thermal constancy) for Neanderthals [is estimated to have been] 27.3°C [81.14°F], while for modern humans it is 28.2°C [82.76°F]. Thus, Neanderthals appear not to have been morphologically well-suited to staying warm in winter conditions without cultural buffering."[9]

Thus, it appears as if *Homo neanderthalensis* became one of the species before *Homo sapiens* that "invented" thermoregulatory culture. It could not have done so without significant evolutionary adaptation of the central nervous system, including the brain. If cultural evolution picked up where this process of genetic evolution left off, it was nevertheless enabled by genetic evolution, which did not simply disappear once humans had invented culture and produced technologies through which to distribute their innate and universal thermoregulatory burdens. The pressure of genetic evolution persists even in our own evolutionary period, in which culture makes the most visible or dramatic impact. But culture requires something else as well: a complex and big brain.

## Evolution creates layers upon layers

In the end, one of the great lessons of social thermoregulation is that genetic (that is, biologically driven) evolution deserves more credit than most cognitively oriented theories or any other culturally based theory gives it. When I first started studying social thermoregulation, it would drive me up a wall how people in my field thought that every human behavior could be activated simply with some form of cognitive priming, such as by reading a few words in a crossword puzzle, as participants in some studies did. Inferences proposed from priming research included asserting that an entire culture might be built on the cognitive priming of basic biological mechanisms that regulate temperature. This is just not the way the world works. Evolution has produced a brain that is much more complex than that. Simply put, we are biologically prepared at birth to participate in culture. Natural selection has produced a hierarchically structured central nervous system, with newer areas of the brain layered over—but not replacing—older areas, making way for ever finer control of various processes, such as the regulation of body temperature.

I first encountered this idea in a 1978 article by the physiologist Evelyn Satinoff, who argued that evolution layers newer functions, such as those governing social interaction, upon earlier functions, such as body regulation.[10] In the article, she invited us to begin by considering an organism able to sense temperature and capable of some form of thermoregulatory behavior. In the course of its evolution, it develops another thermoregulatory behavior, which spurs the development of another integrating system in parallel with the first. Evolution continues, and the reptilian animal changes its posture from the snake's ground-hugging sprawl to something more elevated. This triggers an evolutionary move from ectothermy to endothermy, as the animal develops the capability to produce metabolic heat internally. As this capability develops, so does an additional system to sense and regulate the internal production of heat. Evolution continues further as more advanced forms of movement and thermal regulation are added, along with increased sophistication of the nervous system, which begins to organize many forms of motivated behavior.

Satinoff reasoned that evolution would not undertake the "unnecessarily burdensome" task of creating all-new systems to solve problems that existing systems had already solved. Instead, new systems would be layered upon the old. The result would be a hierarchy of integrators formed atop one another to create a truly "hierarchical organization of thermoregulatory responses for achieving finer and finer tuning of the thermoneutral zone," the range of ambient temperatures within which homeotherms can maintain body temperature with minimal metabolic regulation.

While Satinoff's paper introduced me to this idea of evolutionary layering, Michael Anderson, writing in 2010, came up with even more convincing evidence to support "neural reuse," the idea that newer functions, such as those governing social interactions, make use of older mechanisms, such as thermoregulation.[11] He built his support on a very simple assumption. If newer tasks make use of older brain mechanisms, you can make a simple prediction: the location of an activity in the brain is related to the evolutionary stage in which the mechanism emerged. That is, older mechanisms like thermoregulation are concentrated in the back of the brain, which evolved earlier than the more forward structures, whereas later evolutionary tasks like social interaction are more widely distributed throughout the brain, including into those frontal areas that evolved latest. Although Anderson did not focus on thermoregulation, he provided evidence that it and similar neurological processes are grounded in biology rather than culturally mediated functions of conceptual metaphor.

## Huddling as the foundation of culture

Much happens between infant skin-to-skin care and the diversity of a social network. Some of what happens is culture, which is notoriously hard to define. My former Tilburg University colleague Karel Soudijn and his coauthors counted 128 definitions of *culture* in the period from 1871 to 1987.[12] The famous cultural psychologist Harry Triandis saw some overlap among these various definitions, describing culture as emerging "in adaptive interactions between humans and environments" and consisting "of shared elements [transmitted] across time periods and generations."[13]

Antarctic penguins do the equivalent of all of that, except, perhaps, for transmitting their huddling dynamics to the next generation. (Although who knows?) They spend roughly 38 percent of their time huddling, which allows the individual birds to dial down their metabolism some 16 percent, a conservation of energy that substantially reduces the need to draw on fat tissue.[14] Huddling is much more effective at up-regulating body temperature than shivering, and, for humans, it is a bridge to culture and other ways to regulate temperature, such as reliable indoor heating. Even in homes with good fireplaces or stoves, huddling can still be practiced as a means of staying warm. No doubt it still is in cultures or circumstances that do not offer central heating.

For penguins, a huddle can grow into the hundreds. For people, the huddle has been generally limited to the size of the family or the household, but the point is that the most efficient huddles are more than just two people cuddling. Except for the size of the group, huddling in penguins and in people involves the same two basic principles. First, there must be a level of social organization that promotes cooperation for a common purpose. If penguins or people decide to go their own ways rather than huddle, this method of distributing the thermoregulatory burden becomes unavailable. Second, the heat-generating, heat-conserving mechanics of huddling is pretty much the same. It is a matter of physical proximity that reduces for each individual the peripheral area exposed to the environment.

Obviously, people have developed social thermoregulation well beyond the level of penguins, and that is where the elements of culture start to emerge. For penguins, the size of the group available for huddling is of supreme importance. The more penguins who huddle, the more efficient the distribution of the thermoregulatory burden. But in adult humans, as we have learned, the size of the network is less significant for social thermoregulation than its diversity.

In terms of attachment, something happens to us between infancy and adulthood, which, I confessed in Chapter 5, we "do not yet fully understand." There are a good number of indications that social thermoregulation is important in the unfolding of this process. As we saw in Chapter 5, the desire to thermoregulate socially is closely related to whether you feel comfortable disclosing information to your partner. In addition, researchers from the Karolinska Institute in Sweden have found that being skin-to-skin with your caregiver leads to higher peripheral temperature just as it does in a penguin huddle. Indeed, the temperature from skin-to-skin contact is even higher than that produced by being swaddled in layers of clothing.[15] Finally, research by Ruth Feldman and her colleagues suggests that skin-to-skin contact early in life is important for developing the ability to regulate emotions later in life.[16]

It is true that many mechanisms mediating social thermoregulation, emotion regulation, and culture are not yet clearly explained.

Nevertheless, if you want compelling subjective evidence of the importance of network diversity, walk into a place like Native or Brûlerie des Alpes on a cold day, order a hot cup of coffee, and drink in the gezelligheid. I guarantee you will feel warmer. In part, this feeling can be attributed to the physical temperature of the coffee, but it is also a function of a sense of connection—a feeling of network—with those gathered in a gezellig environment.

## Supersizing the mind

With respect to thermoregulation, social behavior itself does not absolutely distinguish humans from other animals. This is also the case with tool use or culture. Many animal species are capable of moving beyond the limitations of their own internal thermoregulatory mechanisms to maintain homeothermy; they simply use their environment as a tool to extend their own regulation capabilities. For instance, a cold cat may seek a patch of sunlight in which to bask, while an overheated one can seek out sheltered shade. A prairie dog uses burrowing to cope with the extreme summer heat of its grassland habitat as well as with the bitter cold of that same environment in the depths of winter. A typical burrow is quite long, reaching from 16 to 33 feet (5 to 10 meters), and anywhere from 6.6 to nearly 10 feet (2 to 3 meters) in depth. The insulation from the ambient elements is very effective. In wintertime, burrows tend to remain between 41°F and 50°F (5°C and 10°C), and in summer, between 59°F and 77°F (15°C and 25°C). Using such thermoregulatory tools greatly extends the viable range of a prairie dog's habitat.

At least as early as Charles Darwin's 1871 *The Descent of Man*, researchers have described instances of tool use by animals (although naturalists continue to debate just what constitutes a "tool" and, therefore, what animal behaviors can be described as genuine tool use).[17] My conception of a tool is anything that can be purposefully employed to achieve a goal, such as finding protection from excessive heat or cold; it is not limited to an object, such as a stick or a rock, that can be purposefully manipulated. Those who recognize some form of tool use in animal species reject the traditional idea that using tools is one of the highly cognitive behaviors distinguishing humans from other animals. My admittedly broad definition of what constitutes a tool leads us to the observation that humans are not the only animals that use tools to achieve certain thermoregulatory goals.

Among humans, tools and tool use typically have a social dimension; they may be produced by cooperative enterprise, and their proper use may be taught through such social means as schooling, apprenticeship, or other training. Yet tool use is not necessarily a social behavior. Cracking a crab with a stone does not require a committee of macaques. As some animals, including humans, use tools to warm up or cool down, so people and other animal species also off-load their individual thermoregulatory burden through forms of specifically social thermoregulation.

Where there is a clearer break is in language use. Language can transform people's biological and cognitive abilities by simplifying our outer world and giving us unlimited capacities to calculate. Language allows us to better control our emotions, and it also allows us to be committed in long-term relationships (monogamous or not). Language enables us to solve complex problems, including, for example, predicting the weather well in advance.[18] Recall from the first two chapters our studies in which people in warm or cold rooms were asked how close they felt to others. In a study associated with these, we asked people to describe social situations. We found that they used more verbs than adjectives in warm (versus cold) conditions.[19] Verbs, of course, are about action, doing, and making a mark on the world. They are inherently social, whereas adjectives indicate individual properties. Language, the cognitive scientist Andy Clark noted, "supersizes the mind."[20]

## Lakoff and Johnson versus evolution

In Chapter 2, we discussed how language use and culture come together. In their book *Metaphors We Live By*, George Lakoff and Mark Johnson argue that how we think, including about our relationships, is largely shaped by—perhaps even determined by—"conceptual metaphors." This is especially true, they say, of aspects of thermoregulation, offering as proof the proposition that warmth as affection is a universal metaphor.[21] Zóltan Kövecses, in his 2005 book *Metaphor and Culture*, says something similar: "We metaphorically view affection as warmth . . . because of the correlation in our childhood experiences between the loving embrace of our parents and the comforting bodily warmth that accompanies it. This gives us the 'conceptual metaphor' 'affection is warmth.' "[22]

The curious aspect of this theory is that Lakoff and Johnson declare affection-is-warmth a universal metaphor, yet not one that is innate. A cursory survey of several languages seems at first to support the contention that the metaphor is indeed universal. In Palula, an Indo-Aryan language spoken by some 10,000 individuals in the Khyber Pakhtunkhwa province of Pakistan, *táatu híṛu* literally means "a hot/warm heart" and is metaphorically used to express generosity.[23] In the Malayo-Polynesian language subgroup of Austronesian, *senyum yang meng-hangat-kan dada-ku* means "the smile that made me feel warm inside."[24] These languages are very different not only from each other but from the familiar Western languages that use warmth as a metaphor for affection, so we are inclined to agree that affection-is-warmth must be a universal metaphor. Back in 2010, I made this very assumption in my own PhD dissertation, noting that, based on a handful of informants, the metaphor is universal.

I was wrong (again).

My friend and colleague Maria Koptjesvkaja-Tamm, a linguist, has edited a wonderful book about temperature metaphors.[25] She approached the subject much more successfully than I did because she recognized that, to establish universality, you must systematically survey a large sample of languages, preferably from a variety of linguistic families. Maria—her friends call her Masha—surveyed 84 languages, many of which were from the book she edited; she supplemented them with a few examples submitted to her by other linguists.

It turns out that not every language has the same categories for expressing temperature. Some languages have two terms for temperature—warm and cold—some three—warm, hot, and cold—and some even more. In some languages, warm and hot are not the same. Many of these languages use warmth for affection, but other languages do not have this metaphor. Of the 84 languages Masha sampled, affection-is-warmth is absent from 32, and the distinction between warm and hot, which we make in English (for example), is not made in most.

Look at a large enough sample, and you discover that the use of *warm* for affection is found mostly in the Eurasian languages, especially European. Moreover, in some languages in the sample, metaphoric expressions using *warm* for affection appear to be borrowed from other languages. Interestingly, *hot*, used as a metaphor, signifies intensity or danger. *Hot* may be used to describe excessive emotion, great eagerness, enthusiasm, and such passions as sexual desire, anger, actual violence, and dangerous magical powers. More languages use *heat* for anger than use *warmth* for affection. In some languages, *cold* is often used negatively to refer to the opposite of emotional warmth, passion, or enthusiasm, while in other languages and contexts, *cold* refers to calmness, peacefulness, or objective realism or rationality. With respect specifically to the putative universality of affection-is-warmth, even in the case of those languages in which temperature is used as a descriptive metaphor for various emotional states, the specific temperature range expressed by *warm* or *warmth* is not necessarily used for affection. For an overview of the languages that Masha sampled, see the table on the following pages.

| Stock/Family | Language | Affection? | Location | Warm = Hot? |
|---|---|---|---|---|
| Indo-European (IE)/ Germanic | Danish, Dutch, English, German, Swedish, Icelandic | Yes | Europe | No |
| IE/Slavic | Polish, Russian, Serbian, Ukrainian | Yes | Europe | No |
| IE/Baltic | Latvian, Lithuanian | Yes | Europe | No |
| Uralic/Finno-Ugric | Finnish, Hungarian | Yes | Europe | No |
| Uralic/Finno-Ugric | Komi-Permyak (koi) | Yes | Ural Mountains, Russia | No |
| Uralic/Finno-Ugric | Khanty (kca) | Yes | Siberia/Russia | No |
| Altaic/Turkic | Turkish | Yes | Turkey | No |
| Altaic/Turkic | Bashkir | Yes | Bashkortostan, Russia | No |
| Altaic/Mongolian | Khalkha Mongolian | Yes | Mongolia | No |
| Japonic | Japanese | Yes | Japan | No |
| Sino-Tibetan/Sinitic | Mandarin, Cantonese | Yes | China | No |
| Austronesian | Indonesian | Yes | Indonesia | No |
| Algic/Algonquian | Eastern Ojibwe | Yes | Upper MW USA and C Canada | No |
| IE/Italic | Latin | No | Italy | No |
| Isolate | Mapudungun | No | SC Chile and SW Argentina | No |
| Mayan | Yucatec Maya | No | Mexico | No |
| Otomanguean | Zenzontepec Chatino | No | SE Oaxaca, SC Mexico | No |
| Niger-Congo/Gur | Gurenɛ | No | Ghana, Burkina Faso | No |
| IE/Greek | Modern Greek | Yes | Greece | Yes |
| IE/Romance | Italian, French | Yes | Europe | Yes |
| IE/Iranian | Persian | Yes | Iran | Yes |
| IE/Indo-Aryan | Palula | Yes | Pakistan | Yes |
| IE/Indic | Marathi | Yes | India | Yes |

| Stock/Family | Language | Affection? | Location | Warm = Hot? |
|---|---|---|---|---|
| IE/Armenian | Mod. E. Armenian | Yes | Armenia | Yes |
| Uralic/Samoyedic | Nganasan | Yes | Siberia/ NE Russia | Yes |
| IE/Greek | Classical Greek | No | Greece | Yes |
| Sino-Tibetan/Sinitic | Japhung | No | China (Sichuan) | Yes |
| Austronesian/NW Sumatra-Barrier Islands | Nias | No | Indonesia | Yes |
| IE/Germanic-based Creole | Bislama | No | Vanuatu | Yes |
| Austronesian/ Oceanic/Southern Melanesian | Xârâcùù | No | New Caledonia | Yes |
| Austronesian/ Oceanic/N Vanuatu | 18 different languages | No | N Vanuatu | Yes |
| Papuan/Timor- Alor-Pantar | Kamang, Abui | No | The island Alor, SE Indonesia | Yes |
| Papuan, Trans- New-Guinean/ Angan | Menya | No | Papua New Guinea | Yes |
| Non-Pama- Nuyngan/ Gunwinyguan | Dalabon, Bininj, Gun-Wok | No | Australia | Yes |
| Eskimo-Aleut/Inuit | W. Greenlandic | No | Greenland | Yes |
| Athabascan- Eyak-Tlingit/ N Athabaskan | Kuyokon | No | Alaska, USA | Yes |
| Kiowa-Tanoan | Arizona Tewa | No | USA, Arizona | Yes |
| Uto-Aztecan/ Numics | Western Mono | No | USA, Central California | Yes |
| Boran | Bora | No | Peru | Yes |
| Afro-Asiatic/ Semitic | Amharic | No | Ethiopia | Yes |
| Niger-Congo/Kwa | Ewe, Sɛlɛɛ, Akan, Ga, Likpe, Yoruba | No | Ghana, Burkina Faso, Togo, Nigeria | Yes |

| Stock/Family | Language | Affection? | Location | Warm = Hot? |
|---|---|---|---|---|
| Niger-Congo/Gur | Kasem, Buli, Dagari | No | Ghana, Burkina Faso | Yes |
| Niger-Congo/ Adamawa-Ubangi | Gbaya | No | Central African Republic | Yes |
| Niger-Congo/ Atlantic, Mande | Wolof, Mandinka | No | Senegal | Yes |

Thus, we can safely conclude that human beings do not universally perceive social or emotional warmth through the conceptual metaphor of thermal warmth. Obviously, this does not mean that speakers of languages that do not metaphorically equate thermal warmth with emotional/social warmth cannot feel thermal warmth as well as social warmth. They most certainly have the capability. It is, however, clear that they do not use metaphor as a tool to fashion the physical feeling of warmth or cold into a conceptual expression of how warm or cold they feel about their social environment.

Based on Koptjesvkaja-Tamm's data from a reasonably large sample of languages, we know that human cultures diverge with respect to affection-is-warmth. This divergence does not mean that for a Greenlander who happens to speak only West Greenlandic or for an Ethiopian who speaks only Amharic the links associating warmth and affection are broken. They are the links that she and I, along with numerous other researchers—among them John Bowlby, Mary Ainsworth, and Harry Harlow—have observed among attachment in infancy, social thermoregulation, and social attachment in adulthood. It would be absurdly presumptuous indeed to assume that because West Greenlandic and Amharic do not metaphorically equate warmth with affection, speakers of these languages are doomed to faulty attachment and lives without affection. It is also extremely likely that thermoregulation is not the only driver of attachment.

Conceptual metaphor theory (CMT), the view Lakoff and Johnson develop in *Metaphors We Live By*, is thus contradicted by the failure of affection-is-warmth to be universal across all cultures. As we saw in Chapter 2, CMT is also contradicted by the results of the Cyberball video game studies and others discussed in that chapter. Furthermore, for CMT to work, conceptual metaphors must be universal but not innate. Although the expression need not be exclusively linguistic, the metaphor lacking in such a large variety of languages seems to contradict the basic thesis of universality. In short, from Satinoff and Anderson we have learned that the brain simply does not work that way. We also see that social behavior does not work that way, and there is no indication at all in language that CMT is correct. Everything we currently know about homeothermic endotherms tells us that the need for social thermoregulation is both universal and innate, brought about by years of complicated evolutionary development, likely in the layering process that Satinoff described and on which Anderson elaborated.

## Not by hypothalamus alone

The title of this chapter, "Not by Hypothalamus Alone," may strike some readers as an oblique reference to the Gospels of Matthew (4:4) and Luke (4:4), from which comes the popular expression "Man does not live by bread alone." That isn't my intention. It is, rather, a nod to a 2005 book by environmental scientist Peter J. Richerson and anthropologist Robert Boyd, *Not by Genes Alone.*[26]

Richerson and Boyd begin with the observation that humans have evolved to become unique in nature despite being quite similar to other mammals in many ways. Most of the similarities are obvious, but some—such as the use of huddling as a means of social thermoregulation in both penguins and people—are surprising. Richerson and Boyd see the truly defining behavioral differences between people and other animals, specifically social behavior that has created culture, as a phenomenon setting humanity apart from every other organism on our planet.

Richerson and Boyd's subtitle, *How Culture Transformed Human*

*Evolution*, proposes an intriguing idea about culture and human evolution. But let's not neglect the main title. "Not by genes alone" of course implies that human evolution may not be wholly genetic, but we know it is nevertheless highly driven by genetics. Humans have evolved to be more physically active than other species, to have big brains representing a greater proportion of weight than in other animals, and to have the largest cerebral cortex ("higher brain") of all mammals relative to the size of the entire brain. There is a pressing need to cool the brain, and it is very possible that the evolutionary increase in human brain size may have been facilitated by global cooling some 3.2 million years ago.

Increased cognitive capacity surely enhanced the human ability to adapt to virtually every habitat, a trait that also sets us apart. The extremely wide range of climatic adaptation is the result of a combination of biological and cultural evolution. We humans have invented—and continue to invent—a stunningly diverse array of tools and methods to enable subsistence. If each individual human were left to him- or herself, we would not achieve the abundance and variety of inventions we now have at our disposal. This diversity of technology (and therefore the wide range of adaptations) requires extremely varied social networks. Another characteristic that distinguishes humans from other organisms is the size, complexity, and collaborative nature of our societies, civilizations, and cultures.

Thus far in this book, I have repeatedly led up to something that certainly looks like an intersection of biological and cultural evolution. Richerson and Boyd see in humans no sharp demarcation between the two forms of evolution, no absolute road sign that tells us we are leaving the land of biology and entering that of culture. As I have argued against the traditional Cartesian dualism of mind and body, so Richerson and Boyd argue against the traditional dualism of nature versus nurture. They argue that the flow from biological evolution to cultural evolution is not one-way: cultural evolution ultimately affects natural selection and therefore contributes to the shaping of genetically based evolution. Culture is thus neither wholly a product of distinctly human genes nor something that humanity has somehow built on top of natural evolution. Richerson and Boyd describe the unique cultural, social, and technological behavior of humans as an evolutionary adaptation, like walking on two legs. It is critically important to understand that while behavior is a product of evolution, much like bipedal locomotion, the products of behavior— from plumbing to politics—are not simply and directly the output of genetically based evolution. They are, rather, the results of a more complex conversation between genetic and cultural evolution.

What are the results of this conversation? Few humans today live in a predominantly, let alone fully, natural environment—a "state of nature" unmediated by cultural inventions and artifacts. Richerson and Boyd suggest that both the natural and cultural environments shape survival of individuals in human groups through natural selection. Both environments, therefore, affect which genes are transmitted from one generation to another. "Natural" selection is thus influenced not only by natural environments but by cultural ones, too. This claim, which is ultimately a claim that culture is rooted in human biology, is nowhere more relevant than it is in temperature regulation.

Return for a moment to our friends the penguins. Their cooperative social behavior facilitates huddling as a means of distributing the burden of maintaining endothermic homeostasis. Most of us would have no difficulty recognizing this instance of social thermoregulation as a biological evolutionary adaptation. It is easy to view penguin huddling as a social behavior directly "rooted" in genetic evolution because, though social, it does not require a complex culture (such as the work of religion or a nation-state) or impressive technological inventions (such as the internet). Although penguin behavior often appears charmingly human in a Chaplinesque way, penguins do not create parliaments, build cities, invent the internet, or swipe right to find a partner to huddle with on dating apps. It is also unlikely that they will invent a wristband that can heat their bodies up to replace the huddle. The sheer complexity and diversity of human culture so thoroughly mediates the human relationship to the natural environment that it may be very difficult to recognize that the $250 "smart" thermostat controlling your home's heating and air-conditioning system has roots that run deeply and without interruption into the genetic evolutionary imperative we call natural selection.

If you find it challenging to see how even the most modern manifestations of human culture are related to the most primal of biological evolutionary imperatives, it might be even more difficult to accept that these modern manifestations may very well affect our genes. But this is precisely what Richerson and Boyd argue. The cultural practices that invented central heating, including its most recent digital and AI refinements, are not only rooted in biological evolution, they also influence our genetic makeup in an ongoing process of evolution. I admit this claim is complicated. In fact, I don't believe we know enough to prove it yet. Despite these caveats, it seems likely that, over time, living in environments so thoroughly mediated by central heating and other technologically sophisticated means of warming up and cooling down will undoubtedly affect how we relate to one another. And, in turn, how we relate is important to shaping behavior that determines how we pass down our genes from one generation to the next.

## From genes to coffee to technology

Galileo invented the thermometer in 1593. This led to thermometer measurements being used to measure human body temperature, as was first explained by the Venetian physician Santorio Santorio in his 1614 publication of *Ars de Statica Medicina* (*The Art of Statistical Medicine*). Frequently reprinted and reread by generations of physicians, Santorio's book provided body-temperature measurements that physicians could use as a comparison to their patients. It became clear that body temperature was an important part of physiology and had a critical bearing on health. Eventually, physicians reached the conclusion that healthy body temperature was pegged to a very narrow range that had to be maintained regardless of ambient temperature. In this realization we can find the emergence of a cultural, cognitively based understanding of thermoregulation's importance. As the theory of evolution was progressively elaborated over the years following the publication of Charles Darwin's *On the Origin of Species* (1859) and *The Descent of Man, and Selection in Relation to Sex* (1871), it was speculated that endothermy (an animal's ability to maintain a metabolically viable temperature through its own metabolic heat energy rather than relying on ambient heat) may be one of the most important factors organizing evolution.

Fine. But how did we get from sleeping together to keep warm to creating the immense complexity of modern society to keep warm?

Let's return briefly to my two favorite coffee places, Native and Brûlerie des Alpes. Both are gezellig, offering the social and emotional "warmth" that feels (to me) a lot like Christmas with the family and also provides a connection with the larger but still local community. The gezellig climate of the two places is greatly enhanced by the attitude and behavior of the owners, the look and feel of the places, and the kind of people who are attracted to these establishments—a diverse crowd who likely share a desire to experience social warmth and a good cup of coffee or tea.

Let's not forget that both Native and Brûlerie des Alpes are built around those "hot beverages" that *The Big Bang Theory*'s Sheldon Cooper identified as the essential elixir of "social protocol." As we have seen, a number of experiments, including some of my own, confirm that merely holding a hot beverage has observable and measurable effects on socially meaningful emotions and perceptions. If, as Bob Dylan sang, "you don't need a weatherman to know which way the wind blows," neither do you need a psychologist to tell you that a cup of hot coffee or tea can make you feel good—or at least better. You also don't need a psychologist or an environmental scientist or an anthropologist to tell you that the feeling is real—that is, physiological, not just something a "pilot" in your head tells you. Drinking your favorite hot beverage is just satisfying.

The satisfaction may be rooted in a basic biological need for thermoregulation, but let's consider the levels of cultural complexity involved in delivering that satisfaction, especially in the context of a gezellig coffee house. To start, neither tea leaves nor coffee beans are native to the Netherlands or alpine France. Most tea and coffee crops are cultivated in climatic environments far from Europe and often on fairly remote plantations. Large-scale cultivation of both tea and coffee requires a high degree of social, commercial, economic, and technological organization. Furthermore, it requires considerable post-harvest processing, manufacture, packaging, and transportation. Production, transport, marketing, and sales are, in fact, global enterprises possible only with the highest degree of cultural organization and social diversity.

The number of people and organizations involved in keeping places like Native and Brûlerie des Alpes supplied with product—and not just any product, but especially delectable product in multiple varieties—is beyond my ability to estimate. The technologies involved are agricultural, scientific, and commercial, not to mention those used in the extensive communication and transportation across all these industries. All told, satisfying thermoregulatory needs with tea and coffee requires the cooperative functioning of an incredibly diverse social network spanning great distances.

Attachment begins immediately after birth, when the human baby is helpless except for its capacity to attach to its caregivers. That adult humans have adapted to living almost everywhere on the planet is due to our ability to continue forming attachments, not just individual to individual but also through a network in which the division of labor is extensive and complex, with members outsourcing or off-loading various of their needs onto others. This is especially true of thermoregulation. In preindustrial societies, individuals used one another to distribute the thermoregulatory burden through huddling. They may also have sent family members to gather fuel to make a fire or traded something of value in exchange for having someone else gather fuel for them. As cultures became increasingly technological, more-efficient means of thermoregulation were designed and built—more-practical indoor fireplaces, stoves, and, eventually, elaborate central-heating systems. The technology of heating cold air was mastered long before the technology of cooling hot air. It wasn't until 1902 that the U.S. inventor Willis Carrier introduced large-scale electrical air-conditioning; it came into general use by mid-century.

As both heating and cooling technologies developed, supplying fuel became a business. As civilization grew increasingly urbanized, fuel for heating became a more complex business. The competition for fuel became greater and more complex as well. Industry and the evolution of steam-powered manufacturing processes and steam-powered means of transportation competed for fuel. Energy evolved as a major economic "sector," and mining and drilling for fossil fuels evolved into high-stakes enterprises that eventually influenced global geopolitics. Those of a certain age remember the "energy crisis" of the 1970s, when the powerful, mostly Middle Eastern cartel of oil- and natural gas–producing nations, OPEC, held the United States and Europe hostage to an OPEC-created oil shortage. President Jimmy Carter, wearing a thick cardigan sweater, appeared on TV to counsel his country fellows to turn down their thermostats and unscrew unnecessary lightbulbs, all in a bid to conserve precious fossil-fuel energy.

By the end of the 1950s, many of the most advanced nations were applying the most advanced energy technology of the era, nuclear fission, to generate electricity—which led, decades later, to some near-catastrophic results (as in the Three-Mile Island accident near Harrisburg, Pennsylvania, in 1979) or truly catastrophic consequences (including Chernobyl, Ukraine, in 1986, and the Japanese Fukushima Daiichi plant in 2011). More-benign advanced energy technologies followed, including solar and wind-turbine power. On a smaller scale, smart digital technology was applied to heating homes and businesses with the introduction of "smart thermostats" in 2011, which use a machine-learning algorithm to learn users' thermoregulatory habits to regulate interior heating and cooling for efficient, economical comfort. The latest generation of smart thermostats respond to voice command and remote control via internet and WiFi.

As mentioned earlier, for endothermic animals, thermoregulation is second only to oxygen access for sustaining life in the short term. No wonder, then, that it is a vital aspect of infant attachment and that "the discovery of fire" has become a shorthand cliché for the beginning of civilization itself. It has become a status variously enshrined in world mythologies, the most familiar of which is the Greek myth in which the Titan Prometheus steals fire from the heavens for the benefit of humankind. Thermoregulation is at the very core of both biological and cultural evolution. While its importance is primal, issues of thermoregulation have persisted seamlessly, as it were, through both biological and cultural evolution and remain drivers of some of the most pressing geopolitical issues and advanced technologies in contemporary civilization. The evolution has been continuous and without intermission.

## The confluence of genetic and cultural evolution

Without question, social thermoregulation in humans has been extended, elaborated, and ramified by culture and by the product of culture we call technology. But the emergence and flowering of cultural evolution does not end the processes of genetic evolution any more than the confluence of two rivers ends the existence of either river. Genetic evolution is present in human social thermoregulation as its primal and prime mover, as we may infer from the phenomenon of attachment and the mechanism of coregulation (see Chapter 5). Moreover, genetic evolution selected for various morphological adaptations in service to thermoregulation, none of which was sufficient to solve all of the temperature-related challenges.

Yet among the adaptations were some that equipped, prompted, and perhaps even compelled us to embark upon a cultural evolution more capable of solving problems than anatomical adaptations could. Focusing on brain size, the evolutionary coincidence of the enlargement of the brain with the narrowing of the pelvis may well have created the necessity for assisted birth—that is, the rudiments of what we today call society, culture, and civilization. In order to develop fully within the individual, that larger brain, with its especially extensive cortical surface, required more than what the confines of the uterus could offer. This product of genetic evolution demanded the existence and function of cultural evolution, which is at once the product of a large, highly capable brain as well as the mechanism that develops the brain, allowing it to become all that it can be. The question remains open as to how extensively cultural evolution affects more than the individual brain phenotype to instead operate, in the manner of natural selection, on the species' brain genotype.

If we accept the proposition that cultural evolution has a genetic effect, we could speculate endlessly about whether some great evolutionary tipping point will come, at which the effects of biological genetic evolution—truly natural selection—will be overbalanced by the effects of cultural genetic evolution, what we might call cultural selection. For now, however, in our present state of evolution, the evidence argues for the continued presence of biologically driven evolution. For people in technologically developed regions, social thermoregulation relies on social behavior that creates, cultivates, and coordinates a group sufficiently diverse to organize culture, society, and civilization. From these social structures emerge the political, institutional, and technological systems capable of answering each individual's thermoregulatory needs (among many other things). But the prime and primal mover of the behavior of people—and penguins—is the same life-giving, life-preserving necessity: the maintenance of temperature homeostasis.

Yet some may argue that the social thermoregulation found among penguins is foundationally different from the so-called social regulation alleged in humans. Penguins evolved to adapt specifically to a very cold climate. The behavior that produced huddling was absolutely essential to survival. Without it, there would be no penguins today. In contrast, humanity evolved in the warm African savannah. Humans, therefore, had no evolutionary imperative toward a social behavior that organized coordinated means of keeping warm.

This argument is relatively easy to counter. The error here is in overlooking the fact that the savannah gets relatively cold at night. The ambient temperature changes significantly with each and every rotation of the earth.

In August 1973, O. G. Brooke, M. Harris, and Carmencita B. Salvosa published a paper reporting on 12 malnourished Jamaican children, ages 4–16 months, who were studied before and after treatment for their malnourishment.[27] Bear in mind that Jamaica is hot all year round, with very little difference between summer and winter temperatures. Even in the winter, daytime temperatures are 81°F–86°F (27°C–30°C). Nighttime temperatures, however, are 68°F–73°F (20°C–23°C). Brooke and his colleagues observed that "correlations between thermal insulation and rectal temperature in these infants were only significant at night," and that "the main thermoregulatory failure in these children was that they did not increase their heat production in response to cold stress."

Day/night temperature variation provided an imperative to social thermoregulation from the earliest stages of human evolution. Evidence exists that when Neanderthals started carrying infants close to their bodies—thereby thermoregulating them—infant survival increased (even in high-latitude and high-altitude environments).[28] Remember that modern studies show that infants, when swaddled, have lower peripheral temperatures than when they are carried in direct contact with their caregiver's skin. These are all instances of social thermoregulation as an aspect of attachment.

From Satinoff and Anderson, we can conclude that evolution has provided amply for hierarchical innovation in thermoregulation, creating a system of progressively more sophisticated integrators culminating in the master thermostat, the hypothalamus. Yet the evolutionary development of thermoregulation does not end with this structure. It and the system it coordinates are the internal nexus through which many (but not all) of our social interactions are structured. These interactions are founded not upon cognitive social metaphors of physical warmth, but rather on the homeotherm's biological life-and-death imperative to regulate body temperature. What we call society and culture evolved as our brains became more capable of predicting temperature in advance. Indeed, the ability to predict and organize social connections was synchronous with the ability to predict ambient temperature and to organize the means of regulating for it. This is the essence of social thermoregulation—and evolution, not cognition, is at its root.

This said, evolution itself has grown beyond those roots, both demanding and enabling the evolution of a human culture through which social thermoregulation has become a driver of the diversity of civilization itself, spinning off a universe of inventions, some directly and many indirectly related to social thermoregulation. I hope that you, dear reader, are now convinced of the ubiquitous influence of temperature in human life. Such conviction will provide you with the necessary context for the next chapter, in which we explore the modern social applications of imperatives that are at least as old as the emergence of endothermy and its freight of both burden and liberation.

## CHAPTER 7

# Why You Should Sell Your House on a Colder Day

*Temperature in Marketing*

For most of us, our house is our biggest personal financial asset; it represents the single most consequential buying and selling decision we ever make. Earnest buyers research the market and its trends, neighborhoods, local schools, crime statistics, "livability indexes," and an array of other dimensions before taking a leap. Yet fundamentally, a house is neither more nor less than a shelter. As such, it is a tool by which we manage social thermoregulation for ourselves, our families, and our guests.

Today's houses may be very sophisticated in their capability to regulate indoor temperature, but while the reasons buyers choose one house over another may be numerous and varied, the stakes that underlie the regulatory task a house performs are rooted in evolution. Finding a way to create and maintain thermal homeostasis is an issue of survival second only to finding a way to breathe and keep breathing. No matter how many real estate websites we scour, research shows that when temperatures drop, people find houses for sale to be more homelike and warmer, and they are willing to pay more for them. When temperatures drop, social thermoregulation becomes the proverbial high tide that lifts all boats.

Realtors will likely tell you that they don't need psychologists to tell them how to sell a house. According to an article in the Australian online realty magazine *Domain*, "The very best winter display . . . always involves having a great fire as the focal point of a lounge [living] room." The author of the article did seek a comment from a clinical psychologist, who offered, "It's something to do with feeling warm and being reminded of childhood—if you were lucky enough to have a good one—and feeling comfortable and safe and loved." She continued: "When you feel warm, you usually feel good and secure, and when you're cold, all the negatives come up and you start feeling physically uncomfortable . . . not the best circumstances for selling, or buying, a house."[1]

The clinician quoted in the *Domain* article implies that the contrast between a house warmed by a fireplace and the inclemently cold weather outside is an effective driver of sales. Nevertheless, another *Domain* article cites as "one of the most persistent notions" in the real estate market "that selling in spring will achieve the best possible result"; it counters this notion by pointing out that winter provides an opportunity to "capitalize on creating a haven from the outside elements" through such tactics as creating "a warm color palette . . . using earthy shades such as current trends of dusty pink, olive and ochre," which "can warm up even the coolest interior." Lighting should be on the "warm yellow" end of the spectrum rather than the "cool fluorescent," and in the "outdoor winter garden or courtyard," *Domain* suggests a fire pit.[2]

Realtors also tell us that creating a warm feeling in a home offered for sale is not just about literally warming the room with a roaring fire. Indeed, open fireplaces are notoriously inefficient ways of heating an interior space, but the sight and sound of the roaring fireplace certainly feel warm, as do colors associated with warmth. *Domain* suggests that the sense of smell can also be tactically stimulated. "Create a warm, comforting atmosphere that can help put potential property buyers at ease with Vanilla Amber Aroma Crystals. Use real vanilla."[3] Note that the writer quite casually equates warmth with an atmosphere that is "comforting" and helps put prospects "at ease."

## Don't jump to conclusions

A realtor interviewed in *Domain* talked about selling "a big garden terrace that was open to the rugged sea and really blustery weather" in the Australian winter. She turned the heat on but then turned it down during the open house, instead "lighting candles and having mood music playing." She told *Domain* that the house "had previously been on the market for $2.5 million [Australian], but sold . . . for $2.76 million. It was a great outcome—and I think it was down to that [warm] presentation [in winter]."

This suggests that the effect of offering a "warm" indoor environment in the contrasting context of a "cold" outdoor winter environment can be measured in a higher sales price. Perhaps. But, as is usual when we try to apply psychological principles to give hard-and-fast real-world advice, we come to the harsh realization that we live in a multicausal world with a range of possible variables. It is thus important to keep in mind what we would call the "explained variance" in statistics, of which we are usually able to address only a few. In the real estate business, for example, there are built-in variations in buyers' schedules, such as popular vacation periods with more or less free time to look at homes, macroeconomic conditions (economic upswings and downswings), availability of mortgage money, and so on. Moreover, while winter may make the "warmth" of a well-presented house look especially attractive, winter conditions in many regions would require people to traipse through snow and sleet to look at houses and thus discourage them. The value of a given house may be perceived as higher in winter, but a paucity of prospects and competing bids may drive the price down. Such factors predict the selling price of a given house independently from temperature. So, as a disclaimer before we dive into the data, variables other than temperature will have great effect on listing and sales prices.

# Attachment to a home is a thermoregulatory mechanism

Although we should take the multicausal caveats seriously, an awareness of evolutionary and cultural drivers can nevertheless give you an edge, whether you are trying to sell a house or some other product, service, idea, or point of view. There is power in knowing what evolutionary and social thermoregulatory buttons to push with your customer. We should also be mindful of explaining the thermoregulatory linkages in our connection to houses and other "merchandise" one-dimensionally.

History teaches us that humans have continually sought demarcations of space—from caves, alcoves, and grottos to huts and, finally, houses—to shelter themselves from predators and the elements, especially the potentially lethal cold. I believe that the crucial motive of survival has driven people to enhance their houses with functions extending beyond basic survival. The evidence of material culture supports the idea that houses can fulfill a need for affiliation or belonging. The more successfully a house answers this need, the "homier" it is. That is, the house becomes a home or, to use a phrase from attachment theory, is perceived as a "safe haven." Various studies have hypothesized that the cognitive mechanisms by which we perceive a particular spatial demarcation as a home are rooted in the same physiological mechanisms (of attachment and homeostatic thermoregulation) that in early evolutionary and individual development helped shield us from the cold through other people.

Indeed, while early humans and other mammals distributed the thermoregulatory burden across a group, in the course of social evolution the house replaced at least some of these socially borne functions. In one sense, this notion is so obvious as to be self-evident. As cultures developed, the house emerged as a convenient and efficient means of both insulating against the cold and, as indoor heating technologies developed, providing positive methods of generating heat.

To study this, two of my former students, Bram van Acker and Jennifer Pantophlet, and one of their peers, Kayleigh Kerselaers, and I devised a number of studies to test whether houses (pictured in real estate ads) become more attractive when temperatures drop. We wanted to test an aspect of the house-as-a-home idea that was less obvious and more specifically a function of social thermoregulation. We were curious to see if lower ambient temperatures would create a preference for a house in the same way that lower temperatures spark attachment to other people as expressed through renting romance movies[4] or having nostalgic thoughts.[5] Building on the realtors' intuitive belief that "warm" houses are the most salable, we sought to empirically validate how homelike houses enable us to be with others by creating the perception of "feeling at home." Moreover, we sought to investigate whether this affiliation satisfies thermoregulatory needs and whether it would lead to a greater willingness to pay more for an advertised house.

We concluded that the link between temperature and homeyness is mediated by our human motivation to be with others. Then, we created a package of three studies to investigate how lower temperatures elicit what social psychologists Lora E. Park and Jon K. Maner call a "desire for affiliation,"[6] measured by behaviors such as wanting to call or be with friends. In turn, we wanted to know how this need for affiliation motivates a preference for houses that feel homier. To help ensure the accuracy and replicability of our results, we used an online collaborative website called the Open Science Framework to determine our hypothesis and freeze all our decisions before we collected the data, separating our exploratory, hypothesis-generating research from our confirmatory, hypothesis-testing research.

We were confident in our finding that lower temperatures do induce a "need to affiliate." Temperature also predicts how homelike people will perceive a house to be and how much they would be willing to pay for it. This second dimension, because it seems more "objective" (in that it gives a concrete number, like the perception of temperature versus the subjective feeling of "warmer" or "colder"), may support the idea that the desire for homeyness in any particular house offered for sale is linked to the economy of action. Recall that behavioral ecology and, by extension, embodied cognition can inform decisions about opportunities for action weighted against their perceived cost. Once again, a person burdened with a heavy backpack will estimate the angle of a slope to be steeper than a person who sees the same slope but is not wearing a backpack. The "objective" perception of the angle is influenced by the person's estimate of the metabolic cost (in terms of energy) of trekking up the slope. As a hiker contemplating a slope asks herself whether she can afford the metabolic cost of the climb, so a prospective home buyer, though he desires a homey home, asks himself whether he can afford the added cost of its homey features.

We have already seen throughout this book evidence of how central a role thermoregulation plays in the ways we relate to one another. Thermoregulation figures importantly in both genetic and cultural evolution, as well as in the ongoing interactive dialogue between these two evolutionary processes. We have seen how social thermoregulation in particular drives us to organize into and affiliate with more diverse networks, creating cultures, societies, and civilizations. At the heart of it all is the urgent and ongoing need to maintain thermal homeostasis within a very narrow core-temperature band. Doing so is a matter of life and death, although now we have created so many tools that people living in richer countries will not find this struggle to be obviously salient.

The economy of action moves both humans and other mammals to create distributed, and therefore more efficient and predictable, ways of regulating metabolic energy. These steps toward social thermoregulation are far more energy efficient and effective than solitary thermoregulatory behaviors like shivering. As cultural evolution came to play a greater role in human development, we took to wearing clothes and seeking or building shelters to regulate temperature even more predictably. Even as cultural evolution becomes increasingly layered and elaborate, the original evolutionary drivers of thermoregulation continue to operate and make themselves felt. We continue to think of other people as a communal gauge of how well we are protected from life-threatening cold. But the thing is, we think of the relationships we have with houses—and many other products and product brands—in much the same way as we think of our relationships with people. Why is that? This is because some relationships, with animate and inanimate entities alike, play roles in thermoregulation not just figuratively but also physiologically. For others, we simply project our personal relationship models onto our products.

## Not so fast

To a significant degree, we humans have organized our individual relationships as well as our societies, civilizations, and cultures in an effort to keep warm. Just because my colleagues and I believe this is true does not mean it tells the whole story. Our work on the home as a thermoregulatory mechanism consisted of a pilot study and two main studies. We conducted one pilot before we actually ran the main studies.[7] In it, we created hypotheses that were very much in line with the traditional social-cognition principle called priming. We all have a concept of a warm house, so touching something physically warm can be thought to prime us to think that a house we are asked to rate is socially warmer (that is, homier). Based on previous findings in studies by others, this traditional prediction seemed to be a very reasonable hypothesis.

We were wrong (yet again). In our (non-preregistered) pilot study, holding a cold cup tended to prompt people to judge an advertised house as homier. In contrast to our procedure with the pilot study, we then preregistered the two main studies. Preregistration is simply a way of specifying your research plan in advance of gathering data. By preregistering through the Open Science Framework, we separated our preconceived hypothesis from actual data, which enabled us to learn more about the empirical reality of social thermoregulation.

What we found in our pilot study was the very opposite of the traditional social-cognition hypothesis. Therefore, we updated our hypothesis for our two preregistered studies. We also changed our study procedure. To increase the realism of the study, we decided to move outside of the laboratory and university environment. We wanted participants who were more likely to have actual experience with buying a house (in other words, not university students). Partly for convenience as well as for realism, we also decided not to prime participants by having them hold cups, but instead we made use of prevailing warm indoor and cold outdoor temperatures. Finally, while we tested with an advertised house in Study 2, in Study 3 we decided to use each participant's own home, so that we could test the effect of the known associations with one's own house on the perception of homeyness at different temperatures. The hypothesis for both Studies 2 and 3 was that external/ambient coldness would elicit a greater sense of communality toward the house. That is, external/ambient coldness would prompt the participant to judge the house as homier. We also tested the effect of coldness on the perception of a house's attractiveness, the willingness to purchase the house, and its perceived value. What we expected was that each participant's relationship to the house would be mediated by his or her need for affiliation.

The design of the tests included a cold "staying outdoors" condition and a warm "staying indoors" control condition, with a third condition for a cold "going inside." This third condition included a priming step, in which the third group was told that they "would go inside shortly." This was intended to reduce their motivation to self-regulate their warmth, which would have reduced the effects of coldness.

The rationale for this third condition was based on work by Yan Zhang and Jane Risen, psychologists at National University of Singapore and the University of Chicago, respectively. In one of their studies, for instance, they tested 43 participants from the University of Chicago. In the depths of a typically frigid Chicago winter, they asked participants while they were outside or inside whether they

would prefer socially warm activities, such as buying a gift for someone they love, over other temperature-neutral (control) activities,
such as getting a great haircut. When participants were outside, they
preferred the socially warm activities more than the control activities. Inside, however, they preferred the socially warm activities and
control activities nearly equally. The rest of Zhang and Risen's work
also suggested that the goal of socializing (which is activated by
being cold) will be extinguished when the goal is no longer relevant,
as when you are told that you will soon be going indoors.

When I read the original study from Chicago, I did not find the
evidence of this extinguishment to be very strong, but we decided
nevertheless to build upon and test it. We therefore hypothesized in
our own work that the cold outdoor condition (versus both the warm
indoor and the "going inside" conditions) would tend to increase
participants' appraisal of both the communality and attractiveness
of an advertised house, as well as increasing interest in the house and
willingness to purchase it. These effects, we further hypothesized,
would be mediated by the participants' need for affiliation.

What we found was not quite what we hypothesized. The key predictor variable was the actual temperature, exclusively. Regardless
of whether you are told you are going indoors soon and regardless of
need for affiliation, actual temperature predicts the degree to which
people assess a house as more communal, a "safe haven." This, in
turn, predicts both how attractive they find the house and their willingness to pay more for it.

## Houses are people, too

Of course, a house serves a key thermoregulatory function, and this
is important in deciding whether to purchase this house or that one.
But it is also the case that we consumers tend to anthropomorphize
products, starting with those we see as serving us in some meaningful way. In the case of houses, we may view windows as eyes, and we

may speak of the house as happy, somber, dignified, sad, and so on. It may be welcoming or forbidding, emotionally warm or cold. We may say that a house has charm or character, or lacks both. We may speak of the soul of a house—and some people locate that soul quite literally at the hearth. Beyond these straightforward instances of anthropomorphism, we often tend to view houses (and other products) as things that "help" and "protect" us.

In a strictly literal sense, only a fellow creature—either human or, in some cases, animal—has the active, affirmative ability to help or protect. Yet a house is often seen as the personification of the family or of a particular family, such as the "house" of Windsor or Rothschild. A house may even embody the character of a society. The Iroquois, a North American confederacy of six Native American tribes (Mohawk, Onondaga, Oneida, Cayuga, Seneca, and Tuscarora), refer to themselves as Haudenosaunee, or "People of the Longhouse." In the eighteenth century, the *longhouse* was the name of the characteristic Iroquois communal dwelling, built to house 20 or more families. It was also an actual meeting place, a kind of capitol for the Iroquois confederacy. Finally, the "longhouse" was used as a metonym (or metaphorical expression) for the confederation itself as well as its geographical place in the world, stretching from the land of the Mohawks at the northeast end of the "longhouse" in the Adirondack Mountains to that of the Seneca at the southwest end, adjacent to Lake Ontario and the Genesee River. The people of the metonymical longhouse helped one another, creating conditions of warmth, security, and protection from danger even as the physical longhouse dwelling provided the very same things.

In building or buying a house, we recapitulate on an intimate scale many of the benefits our town, nation, society, network, or civilization affords us. A house is a social structure that embodies social functions ranging from the primal thermoregulation created between infant and caregiver to the more elaborate connections that join us to the diverse network of society. A homeless person, on the other hand, is exposed to dangers including temperature extremes and is effectively separated from society, much like the penguin who wanders away from the all-embracing and life-preserving huddle in Werner Herzog's documentary.

As we anthropomorphize our homes, so do we anthropomorphize many products and objects as thermoregulating tools. In some cases, we even project onto objects that seem to have nothing to do with keeping us warm. In 2014, I conducted a study with my student Janneke Janssen (whose idea was the impetus for the study) and colleague and friend Jim Coan, whose work inspires much of my own. Together, we explored how consumers maintain "warm, trusting relationships" with product brands.[8]

"Communal brands" are those associated with what marketers call a "brand community," a community based on shared attachment to a product, logo, or brand. Marketers believe that communal brands are connected with individual identity and culture. Our group proceeded to create a set of studies to test the hypothesis that just thinking of communal brands would increase your perceived temperature. We tested a pretty large sample (2,552 people) from Amazon Mechanical Turk, a website very popular among psychologists doing research. In five studies, we found that thinking about positively perceived communal brands leads people to estimate ambient temperature as higher. What is perhaps more curious (and very much in line with the studies on houses) is that some exploratory analyses we conducted in this set of studies suggested that the increased temperature perception drives people's willingness to purchase the brand as well as the amount of money people are willing to pay for it.

Marketing directors use this knowledge, too. Coca-Cola is a classic global communal brand. As a beverage, it is typically served ice-cold and promises welcome refreshment in the heat of summer. Nevertheless, one of the most familiar modern advertisements for Coca-Cola shows large, illuminated red delivery trucks emblazoned with the iconic white calligraphic Coca-Cola logo making their way through a winter landscape. It is an image that offers relief from the bleak cold of winter in the form of warm lights adorning the trucks, evoking feelings of Christmas cheer. Presumably, the brand's marketers, legendary for their skill, were seeking to create or reinforce a trusting, warm relationship between the brand and consumers. I do not know if those marketers were aware that the emotionally warm feelings associated with their brand are reflected in physically warm feelings, but it's true. Our study found that when consumers deal with brands they perceive as trustworthy, they estimate ambient temperature as warmer than they otherwise would. That is, brands with which consumers have close relationships instill an increased sense of physical warmth.

With its large sample collated from the five studies, the article we published in 2014 provided greater confidence in our hypotheses—and did so with more precision. I was skeptical that people's subjective temperature feelings (that is, feeling warm or not) would be reliably influenced by communality, and we turned out to be right. Subjective temperature was not influenced, but perceived ambient temperature was.

The thermoregulatory effect of brands is consistent with the long-standing perception that brands fulfill relationship functions; moreover, they do so in ways customarily reserved for person-to-person relationships. This suggests that mechanisms associated with attachment and coregulation in interpersonal relationships are also active in our relationships with brands. Studies of people who collect Hallmark and Coca-Cola products and advertising artifacts reveal that people maintain fully, even profoundly, affectionate relationships with favorite brands. One study suggested that to ardent collectors, "Hallmark is like a lover" and that their relationship with the collected brand-related items is an enduring and dependent one, progressing in phases analogous to "dating, seducing, and tying the knot."[9]

Nor is such pronounced anthropomorphism exclusive to enthusiastic collectors. Many people direct toward products the kind and degree of affectionate feelings associated with social attachments.

Aaron Ahuvia, professor of marketing at the College of Business of the University of Michigan–Dearborn, is a noted expert on "brand love." In one study, he interviewed 69 participants, of whom just two asserted that love is strictly reserved for human beings, to the exclusion of objects or products. Asked if love toward consumer products could be described as "true" and "real," 72 percent of the respondents answered yes.[10] Studies by others have reinforced the findings of this very small-scale study. In 1988, Terence A. Shimp and Thomas J. Madden, both of the University of South Carolina, developed a model of "consumer-object love,"[11] which captured components traditionally thought to be exclusive to love between people, and Marsha L. Richins identified a set of descriptors of emotional attachments to brands, including "love," which is measured by the degree of "warm-heartedness" felt toward a product.[12]

Consistent with product anthropomorphism is brand personality, by which people evaluate their favored brands using terms typically applied to personal attributes. A brand may be judged as "cool" (in a fashion sense), innovative, athletic, authentic, practical (in a realistic sense), stylish, or trustworthy. We all judge both brands and people, often on the spur of the moment. Psychologists have shown that it is useful, even adaptive, to make such snap judgments, because we want to know whether someone wants to hurt us and is capable of doing so. This is related to what psychologists have defined as the "Big Two" dimensions of how we perceive other people, called "communion" and "agency." Communion, or emotional warmth, incorporates such personality attributes as trustworthiness, helpfulness, honesty, and cooperativeness, while agency focuses more on competence and includes such qualities as efficiency, persistence, and energy. All of these human attributes are sometimes projected upon brands.

Overall, in five studies, we found that judging a brand to be communal also leads participants to estimate temperatures as higher. We think that brands answer to similar psychological concerns related to the economy of action; that is, when it is cold, we probably turn to psychologically warmer brands to satisfy some inner affiliative needs. Our preference for trustworthy brands is thus also rooted in the regulation of body temperature to some extent, just as our attachment to trustworthy members of a diverse social network is rooted in early attachment to reliable caregivers. And brands seem capable of giving some stability to our otherwise changing world, thereby helping us better predict the future. It is similar to checking the weather report to find out what the weather will be like during the next 10 days.

Psychologists are frequently asked to apply their theoretical ideas to produce practical and predictable results. Therefore, to create successful communal merchandise, I suggest that marketers ask: How can products and brands be designed to sell better by eliciting the perception of warmth and thus trustworthiness? How can a product or brand be made to elicit the perception of positive communality? The first and simplest answer I'd offer is not just to create a trustworthy brand, but also to be trustworthy. The fuller, more complicated answer is that, at present, the link between psychological theory and application is not straightforward. Contrary to a recently created belief from my discipline, now substantially debunked, standing in an upright pose is not sufficient to achieve a winning job interview.[13] We can, however, learn from highly powered studies testing theoretical predictions and make rational educated guesses from the resulting data.

Marketers must try to understand the necessary preconditions for solid science and then know what is relevant to measure. In the case of using social thermoregulation to predict the commercial viability of a brand or product, the relevant measurement is the perception of ambient temperature, expressed objectively in actual degrees, rather than the subjective expression of feeling warmer or colder. If you are a marketer looking for a practical application, please understand that the information is rooted in a relatively complex hierarchical evolutionary layering of neurological regulators, where temperature will explain only a small amount of the variance. This variable can give you an edge, but only after you have thought through all the other variables relevant to selling your product.

How the choices of the majority influence the choices of the individual is a powerful variable that can be tested independently of temperature. For instance, in 2001, Sushil Bikhchandani and Sunil Sharma reviewed research on "herd behavior" in financial markets, working from the hypothesis that conformity is based on the common assumption that if the majority chooses an option, it must be good.[14] The countervailing behavior—nonconformity—assigns greater value to the minority evaluation, concluding that the option chosen by the few is the more valuable or valid.

As Wayne D. Hoyer and Deborah J. MacInnis observed in their college textbook *Consumer Behavior*, marketers use both conformity and nonconformity to promote products.[15] Some brands tout their majority appeal while others promote exclusivity, the special discernment of an elite minority. Given all that we know about thermoregulation, conformity practically cries out to be linked to social thermoregulation. Alas, there are no reliable studies on the topic, but for the sake of investors, bettors, and all team sports alike (not to mention social psychology), it is absolutely imperative that these studies enter the literature soon.[16]

## Thermoregulation and emotional advertising

Until we more fully understand the physiological and psychological mechanisms involved in temperature and decision making, attempting to explain why ambient cold prompts certain decisions and ambient warmth prompts others will remain largely speculative and tentative. Yet, even before the business of selling products to people became increasingly specialized and sophisticated in such fields as marketing, advertising, and public relations, all of which now draw on sociology and psychology, successful salespeople had already intuited links between emotions and temperature to help them promote their wares.

In a 2017 article, Pascal Bruno and his colleagues followed the popular metaphor idea and observed that people often use warmth with love and coldness with fear.[17] Nevertheless, they went further, as they also observed that professionals in the business world use emotionally warm and cold appeals in advertising. This was something they sought to investigate. Are there specific conditions under which emotional warmth, or emotional coldness, appears more or less effective?

The researchers turned to ideas similar to those we have introduced when they evoked temperature homeostasis—internal optimal balance within a biological system—as a possible force that shapes social behavior. Through laboratory experiments and field data, they tested the hypothesis that the effects of certain emotional stimuli depend not only on physical temperatures but also on their role in thermoregulating toward homeostasis. They concluded that consumers who are physically cold (operating below their homeostatic optimum) perceive emotionally "cold" stimuli less favorably than emotionally "warm" stimuli. The researchers' explanation was that the cold stimuli push participants further from the homeostatic optimum. In contrast, when consumers are physically hot (operating above their homeostatic optimum), they perceive emotionally warm stimuli less favorably for the same reason. That is, they are pushed further from homeostasis. When consumers are at their homeostatic optimum—like Goldilocks, feeling neither too cold nor too warm—they perceive both emotionally warm and emotionally cold stimuli with a similar degree of favorability.

Understanding that temperature and emotion affect homeostasis similarly, Bruno and colleagues cited the work of cognitive neuroscientists Antonio and Hanna Damasio on how emotions can elicit changes in the body as if the body really responds to temperature changes in the environment.[18]

Bruno and his colleagues recruited 461 students (49.7 percent women) and divided them into two groups, putting one in a room with a comfortable temperature of about 86°F (30°C) and the other in a room with a cold temperature of approximately 57.2°F (14°C). The participants in both temperature conditions were then exposed to one of 11 advertisements representing either emotional warmth or cold. Before participants were exposed to the advertisements, they recorded their mood by responding to questions on a 7-point scale ranging from 1 ("not at all") to 7 ("very much")—a Likert-type scale standard in much psychological research. Measuring participant emotion in this way was intended as a control variable to rule out potential alternative explanations for increased or decreased liking of the emotionally warm or cold advertisements.

The participants were asked to rate their assigned advertisement on six 7-point scales: liking, interest, convincingness, appeal, the ad's potential to be remembered, and its effectiveness (1 = "not at all," 7 = "very high"). They also noted their likelihood to purchase based on the ad (from 1 = "not at all," to 7 = "very high"). Finally, the researchers asked the participants to record their perception of the room's physical temperature on a 7-point scale (1 = "very cold," 7 = "very warm"). For emotional temperature, they indicated to what extent looking at the ad made them feel warm (1 = "not at all," 7 = "very much"). The results showed that the emotional temperature condition elicited by viewing either an emotionally warm or emotionally cold ad significantly affected perception of physical temperature. After exposure to emotionally warm ads, participants felt the room's physical temperature was significantly higher. After exposure to emotionally cold ads, they felt it was lower.

The researchers also found that participants who felt physically cold perceived emotionally cold advertising less favorably than emotionally warm ads. This affected both their attitude directly toward the ad and their likelihood to purchase the advertised product. In contrast, those in the comfortable physical-temperature condition, within homeostatic optimum, displayed no difference in their responses to emotionally warm versus emotionally cold ads, with respect to both the ad itself and their likelihood to purchase the advertised merchandise. These results were consistent with the researchers' prediction that participants at homeostatic optimum would not respond differently to emotionally warm and emotionally cold ads.

In another study reported in the same article, Bruno and his colleagues ventured further from the homeostatic optimum, testing effects at the optimum level as well as at temperatures uncomfortably below and above it. They tested their lab results with a field experiment that used television spots (rather than print ads) in an outdoor setting. This allowed observation of responses to ads across various outdoor temperature levels. They found that, at low temperatures, the emotional warmth of a spot increased its persuasiveness. At medium outdoor temperatures, closer to optimum, the emotional temperature of the spot had no significant effect on its persuasiveness. This finding was consistent with the indoor lab studies. Finally, high outdoor temperatures, above the optimum, enhanced the perceived persuasiveness of emotionally cold ads while decreasing that of emotionally warm ones.

## The thermoregulatory cost of regret

We have seen throughout this chapter examples of how social thermoregulation relates to the perception of the social warmth of houses, products, and advertisements and how this influences people's preferences. One way or another, all of these studies ultimately involve financial decision making.

All financial decisions are economic decisions, but not all economic decisions are financial. At the level of social thermoregulation, all of the decisions the foregoing studies investigate are also rooted in the economy of action—that is, in how we choose to invest our scarce metabolic resources. If you happen to be a penguin, you cannot exchange cash for physical warmth or social warmth. You can, however, "choose" to invest your metabolic resources in social capital, namely the organized proximity of a group of other penguins in a huddle. Modern humans can and very often do invest their scarce financial resources to distribute their thermoregulatory metabolic burden across a diverse social network. The drivers of such financial decisions remain rooted in the physiological imperatives of the economy of action. Indeed, finances—money spent and not spent—are a convenient objective metric of how we invest our metabolic resources. And that, of course, is at the heart of social thermoregulation.

Persuasion is one phase of the journey toward a decision to act or not to act. But that decision does not end the journey. We may be pleased with our decision or we may regret it. Marketers and salespeople are very familiar with "buyer's remorse," the post-purchase feeling we may have regretting the purchase of a product now perceived as extravagant, too costly, unnecessary, or in some other way inadequate and therefore a bad bargain. Psychologists see buyer's remorse as a subset of the more general emotion of "action regret," which follows from a behavior that you did not want to engage in or now feels like a mistake or bad choice. It is also possible to experience "inaction regret," which is regret that follows the failure to act on something that you wanted to act on or now realize that you should have acted on. Social thermoregulation plays a role in the aftermath of decision making just as it does in the run-up to the decision.

A 2017 article by Jeff Rotman and others suggests that the experience of action regret produces a change in both emotional and physical warmth.[19] One of their experiments revealed that advertisements manipulated for temperature—for example, an ad depicting a cold climate—ameliorated the emotional effects of action regret. They tested the idea that getting rid of the effects of action regret does not necessarily require a physical product such as a cold drink. Just as effective for amelioration may be simply imagining the experience of cold. For instance, advertisements featuring temperature-related attributes can reduce regret stemming from an action. The proposal behind this experiment was that participants experiencing regret who view an advertisement for a vacation in the cold, such as a ski holiday, will feel less regret than those who view an advertisement for a warm vacation, such as on a sunny Caribbean beach.

The experimenters conducted an online experiment in which they analyzed data from 119 participants who were provided information about a fictitious pharma stock, Verap Pharmaceuticals, at a current share price of $2.50. The participants were given the option of investing (action) or not investing (inaction) in the stock with money that they received from the experimenters. So that participants would feel regret coming from an action, participants who chose to invest their money were informed by the researchers that their stock had dropped to $1.25. Those participants who did nothing and therefore kept their money were told that the stock had risen to $3.75. Of the sample Rotman and colleagues tested, more than half (53.8 percent) chose to invest and experienced action regret; 46.2 percent chose not to invest and experienced inaction regret.

After the regret-inducing phase of the experiment, the participants were shown an ad for a cruise vacation, and were then asked to imagine themselves on the cruise and to provide an estimate of the temperature they would experience. Half viewed an ad for a cruise in a cold environment (Alaska) and half saw an ad for one in a warm environment (the Caribbean). After seeing the ad, participants were asked to complete a four-item, 7-point Likert decision-regret scale regarding the stock decision they had made. The scale ran from "strongly disagree" (1) to "strongly agree" (7) and included such statements as "I regret the choice I made" and "I should have chosen differently than the one I decided." (In addition, the experimenters collected the usual demographic variables and statements concerning happiness. None of these variables influenced the results.)

Unsurprisingly, participants who viewed the Caribbean ad estimated the temperature there to be higher (a median of 67.38°F [19.65°C]) than people who saw the Alaska-vacation ad (median,

30.18°F [−1.01°C]). As the researchers had predicted, among those experiencing action regret, those who then observed the cold Alaskan-cruise advertisement felt less regret than those who viewed the ad for the warm Caribbean cruise. On the 7-point Likert scale, the regret median for Alaska was 4.53; for the Caribbean, it was 5.21. Interestingly, those experiencing inaction regret felt less regret after observing the warm Caribbean ad compared to those who saw the Alaska ad.

While I would have preferred the sample size to be larger, these study results suggest that imagining a certain temperature scenario does indeed affect your emotional experiences. Specifically, it can reduce the degree of decision regret. But the relationship between an imagined temperature condition and the degree of decision regret is not simple. In the case of action regret, the study suggests that imagining a cool ambient temperature reduces the reported level of regret and makes the person feel better. But in the case of inaction regret, imagining a warm ambient temperature reduces the reported regret. I suggest that this difference should prompt us to look beyond temperature per se and interpret the significance of temperature in the context of homeostasis, for the point is that both regretted action and regretted inaction make an impact on homeostasis.

Severe threats to homeostasis can produce serious debility, sickness, and even death. Relatively minor threats, such as those introduced in this experiment, are not nearly so dire in potential consequence, but they can still make us feel bad. And that is important information for marketers. In an episode of the popular television series *Mad Men*, the show's troubled hero, Madison Avenue genius Don Draper, defends the often-attacked ethical and social value of his profession by saying, "I think people buy things to make themselves feel better." I might have put it a little differently—"I think people buy things to help them maintain homeostasis"—but then again, I'm not a marketer.

Nevertheless, we psychologists are often asked to help marketers devise messaging strategies to promote the "feel good"—that is, the homeostatic—benefits of their products. Such requests are valid, and science can be of real service in fulfilling them. The rub, however, is that in terms of both biology and culture, humans are incredibly complex organisms. This means that the results of experiments notwithstanding, there are no psychological silver bullets. Inducing thoughts of a cold place can very likely reduce the discomforts of regret but perhaps only if the regret is for an action taken. For regret concerning inaction, warm thoughts may possibly help promote a product, but the question remains: How is the marketer to know whether her target consumer regrets an action, regrets an inaction, or regrets nothing at all?

Comparing two more studies gives us another window into both the potential and the limits of applying the psychological insights of social thermoregulation to "practical" or "real-world" applications such as marketing.

In a series of five field and laboratory studies, Yonat Zwebner and her colleagues explored what they called a "temperature-premium effect." This is the apparently straightforward proposition that if you expose people to physical warmth, they are then more poised to think about emotional warmth, which in turn elicits positive reactions and increases how positively people evaluate a product.[20]

Let's look at just the first of the five studies. The researchers wanted to know whether temperature affects product evaluation in the real world, which is far messier than the laboratory. Accordingly, they analyzed data gleaned from a major Israel-based price-comparison online-shopping-portal website. The site organized merchandise offerings into broad product domains subdivided into categories and subcategories. Users may search for entire categories or for specific products and can compare the prices offered by multiple sellers. Clicking a To-Purchase button for a seller takes the user to that merchant's website. The number of To-Purchase clicks in a given product category generates for researchers a convenient measure of shoppers' intentions. Zwebner and her colleagues analyzed 24 months of click data—6,364,239 clicks, from September 2010 to

August 2012—in eight categories. They collated with this data daily temperatures to determine how this affected intention to purchase within each of the eight categories. Calculating the average temperature for each day, they performed three regressions to find out if physical warmth increases the intention to purchase.

As the researchers had predicted, temperature has a positive effect on intention to purchase; however, this "temperature-premium effect" proved to be nonlinear. That is, as the temperature increased, its positive effect on the intention to purchase diminished. Yet the temperature-premium effect looked to be quite real, persisting even after the researchers controlled for the effect of holidays and seasons.

Another study, by Peter Kolb and others, took the shopping experience out of the online environment and into the realm of the brick-and-mortar store, focusing not on consumers' behavior but on that of customer-service, or customer-oriented, employees and salespeople.[21] The study consisted of two experiments. The first, conducted in a laboratory, demonstrated that participants—who were university students and not professional salespeople—in a room with low temperature exhibited more customer-oriented behavior and offered customers bigger discounts than participants in high-temperature rooms. The same results were produced even when the cooler and warmer rooms were regulated within zones of thermal comfort.

In the second experiment, researchers tested the effects of alternative means of manipulating temperature to increase the customer orientation of salespeople and customer-service personnel. This time, the researchers tested a sample of 126 service and sales employees, not students. They used "semantic priming" as an alternative to physical temperature manipulation. Participants were asked to solve a word-search puzzle, the object of which was to find 12 given words in a matrix of letters. The participants were randomly assigned to three conditions, all of which included the same six temperature-neutral words. In the warm and the cold conditions, six temperature-related words were added.

After being primed by working the puzzle, the participants' degree of customer orientation was measured with the "Selling Orientation–Customer Orientation Scale." The version used in this experiment was modified to make it relevant to both salespeople and customer-service personnel. Participants rated each test item on a Likert-type scale ranging from 1 (never) to 9 (always). Participants who had been primed by the puzzle containing six cold words demonstrated a higher degree of customer orientation than those primed with either warm or neutral words. From this result, the researchers concluded that, regardless of actual ambient temperature, the activation of the concept of coldness (versus warmth) through semantic priming led experienced sales and customer-service employees to score higher in self-reported customer orientation.

These two experiments are especially interesting because they so freely cross the Cartesian brain-body barrier by demonstrating that effects produced by physical temperature can also be produced by thoughts primed with mere temperature-related words. At the same time, the experiments demonstrate the limitations of pulling practical applications from psychological experimentation. The researchers themselves frankly point out that they do not recommend priming salespeople and service providers at the start of each workday with things like temperature-related word puzzles. All they offer is the rather tepid suggestion that employees drink cold water and wear comfortable clothes to avoid feeling too warm on the sales floor.

Far more interesting to me is what some may see as a real dilemma when it comes to using actual temperature to improve sales performance. The study by Zwebner and colleagues suggests that warmer temperatures prompt consumers to up-value products and make more purchases. Kolb and colleagues, however, suggest that lower temperatures lead customer-service and sales personnel to treat customers better and therefore create a customer-service experience more likely to prompt sales and increase customer satisfaction.

Well, what's a store manager to do? Turn up the thermostat to prompt customers to buy but thereby risk suboptimal customer orientation from salespeople? Or turn the thermostat down, thereby eliciting top performance from salespeople who now, however, must confront customers who may be too cold to make a purchase?

In truth, I don't regard this as prima facie proof that psychologists cannot be relied on to make practical recommendations. Ultimately, the dilemma implied by these two results is a practical dilemma only if we make the mistaken assumption that we are dealing with a simple cause and effect: ambient temperature and human behavior. In fact, the issue here is homeostatic regulation. Purchasing a product fulfills a short-term need, which has a thermoregulatory component. But customer service is more closely related to a longer-term social investment in relationships, which relates more directly and deeply to homeostasis. Customer service, then, is a social act rooted in attachment, the distribution of the life-preserving thermoregulatory burden on others whose presence and support are predictable.

The lessons psychology provides here may be more profound than making a sale or improving customer service. Sales and service personnel are people whose livelihood depends on successfully creating trusting relationships with others. This is something the healthy among us not only strive to do but also succeed in doing. It is the basis of a viable society and a productive civilization. Yet it is even more than that. Homeostasis is essential to survival, and, as will see in the next chapter, thermoregulation is a key to understanding— and perhaps curing—some of humanity's most devastating mental and physical ills, too.

## CHAPTER 8

# From Depression to Cancer

## *Temperature as Treatment*

**Ana, an elderly woman** from São Paulo, Brazil, struggled with depression for many years. Compounding her suffering was another struggle: a perpetual sensation of physical coldness. Even in the warmest weather, Ana felt chilly, and when the weather got colder, she was far more uncomfortable than most. She took to dressing in layers upon layers. She saw a psychotherapist for her depression and, from her, picked up the habit of drinking hot tea; she found that it gave her a modicum of relief from the persistent chill. She also discovered that the tea somewhat lifted her spirits, however fleetingly. She did not, however, see any connection between her depression and her chronic feeling of being cold.

Likely most observers would see no significant connection, but recent research suggests a possible link between depression (major depressive disorders) and thermoregulation. There is experimental evidence suggesting that manipulating our inner thermostat may prove to be an effective therapy for depression. For example, in a 2013 study, each of 16 severely depressed adults reported an impressive improvement in their mood after just one session of whole-body heating with infrared lamps. The improvement was still apparent six weeks after the treatment and was significant enough to shift some of the patients from severe to moderate depression.[1]

That study's sample was very small, which is understandable given the inherent difficulty in identifying and recruiting severely depressed individuals to take part in a psychological study. Still, the smaller the study, the more tenuous the conclusions that can be drawn from it. Most of the studies described in this chapter are relatively small and their implications therefore more suggestive than conclusive. This said, we are engaged in a larger discussion of thermoregulation's crucial importance to survival and health, and we have repeatedly found evidence that social thermoregulation is implicated in functioning more generally.

That thermoregulation is important for survival is not at all controversial. People die all the time from extreme heat or cold, and often in ways that can be avoided. France, where I live, is still haunted by memories of a heatwave in August 2003, in which an estimated 15,000 people died from extreme heat. A recent report from New Zealand suggests that 1,600 deaths each winter are attributed to houses that are too cold.[2] This chapter goes further. I believe we have accumulated ample evidence to suggest that social thermoregulation plays a major role in our general health, including in many ways that are less obvious than dying from the summer heat or succumbing to hypothermia in the winter.

In this chapter, we will encounter practices relating to social thermoregulation that have the potential for reducing the risk of diseases as varied as depression, diabetes, and cancer. Diverse, distinct, and seemingly unrelated as these diseases are, they appear to share one salient link: brown adipose tissue, or BAT, as we have come to call it. Recall from earlier chapters that BAT is related to thermoregulation and, via the mechanism of attachment, to *social* thermoregulation. While having more BAT is viewed in the medical profession as a potential cure for obesity in the short term, studies suggest something different that is likely more complex, longer term, and of even greater consequence.

We know that social thermoregulation, in penguins and humans alike, is necessary to survival. The mechanism of attachment, by which infants off-load much of their life-sustaining thermoregulatory burden onto their hot mamas, leads to the increasingly more complex social thermoregulation that drives human integration into diverse social networks. Can it be, then, that the proximity of others—society—lowers our risk of disease?

## Thermoregulation and its links to our health

Before we get too deep, let's brush up on some basic terms. *Hyperthermia* is what happens when the body gets too hot, that is, when it generates more heat than it can lose. *Hypothermia* is what happens when the body gets too cold, that is, when it loses more heat than it can generate. (Feeling cold, by the way, is always the absence of heat, since heat energy always transfers from hotter to colder places.)

It is quite acceptable to say that thermoregulation is related to health and vice versa. Thermoregulation was already a focus for Hippocrates, who counseled that "whoever would study medicine, must . . . study the warm and the cold winds, both those which are common to every country and those peculiar to a particular locality." I would hazard an unscientific guess that few are the children who, seeking a day off from school, have not complained to their mother of sickness, only to have her apply her hand to their forehead, conclude the absence of fever, and rouse the youngster from bed. Fever has long been used as something of an objective measure of illness or bodily disorder.

And this is for good reason. Human body temperatures above 102.2°F (39°C) are typically accompanied by exhaustion, malaise, and a general feeling of being ill. At 104°F (40°C), there are more dire symptoms, such as fainting, dehydration, vomiting, and dizziness. At this temperature threshold, the fever begins to become life-threatening. A body temperature of 105.8°F (41°C) is a true medical emergency, and at 107.6°F (42°C), delirium, convulsions, and coma may occur. An elevation of just 1.8 Fahrenheit degrees more, to 109.4°F (43°C), often brings death.

Excessive ambient heat can be hazardous when the body absorbs more heat than it can dissipate. Hyperthermia can be caused by high ambient temperatures and becomes life-threatening when body temperature hits 104°F (40°C).[3] It can also be caused by intense exercise. Working at maximal intensity, skeletal muscles can increase their energy consumption 20-fold. Consider that the human body's metabolic efficiency is about 25 percent, and it becomes apparent that much of the energy is converted not to muscular work but to heat, which is transferred from muscle to blood, the circulation of which raises core body temperature. Heat stroke, associated with a body temperature at or above 104°F (40°C), may be produced by excessive ambient temperature, extreme exertion, or a combination of the two.

As we noted in Chapter 3, most homeothermic animals, especially humans, are more tolerant of downward deviations from "normal" body temperature than they are of upward deviations. This difference explains why so many more people die when the mercury goes up than when it goes down. The asymmetry is quite dramatic. A fever of 102.2°F (39°C) is associated with debilitating sickness, whereas hypothermia between 89.6°F and 95.0°F (32°C and 35°C) is considered mild (producing mainly shivering) and between 82.4°F and 89.6°F (28°C and 32°C) is classified as moderate (inducing drowsiness without shivering). The extremes of these two states represent a departure of between 3.6 and 16.2 Fahrenheit degrees (2 and 9 Celsius degrees) from normal. Severe hypothermia (68.0°F–82.4°F, 20°C–28°C) brings unconsciousness, and below this point is death. The authors of "Accidental Hypothermia," an interesting study produced in 1988 for the U.S. Army Research Institute of Environmental Medicine, dramatized the potential impact of prolonged exposure to cold conditions by observing that the Twelfth Division of Napoleon's Grande Armée marched into the disastrous Russian campaign of 1812 with 12,000 men and returned to France with just 350, the rest having died, mainly of hypothermia.[4]

The body, however, can be influenced in much subtler ways. Some surgical anesthetic agents can, in rare instances, cause malignant hyperthermia when the drug stimulates excessive release of calcium from the sarcoplasmic reticulum, a membranous structure within muscle cells that is also involved in shivering and nonshivering thermogenesis. The calcium release causes severe muscle hypermetabolism, which generates heat. When calcium is released in excessive quantity, it generates heat at levels higher than the body can get rid of. In these admittedly rare cases, malignant hyperthermia occurs. Neuroleptic, or antipsychotic, medications, another category of drugs affecting the nervous system, can trigger neuroleptic malignant syndrome, which induces high fever and body temperatures often exceeding 105.8°F (41°C).

Thermoregulation, health, and rudimentary connections to our emotional lives can be observed in very basic biological mechanisms. Hyperthyroidism, a condition due to excessive production of thyroid hormone, can result in hyperthermia due to thyrotoxic crisis, sometimes called a thyroid storm, in which fever may reach 105.8°F (41°C). Hyperthyroidism can be precipitated by stress, including emotional stress. Long-term stress can even be a cause of autoimmune thyroiditis, which causes numerous problems, including sweating, increased heart rate, and difficulty sleeping.

Differences among people can make an individual more or less susceptible to hypothermia. Your thermoregulation can be impaired, for example, by conditions that directly affect how your hypothalamus functions. (Recall our thermoregulation-integrator discussion.) Among these conditions are traumatic injuries to the brain, pathological brain lesions, and diseases such as Parkinson's and Hodgkin's. Spinal cord injury, such as severing of the cord, can cause victims to become poikilothermic and unable to regulate their own temperature. Neuropathies and diabetes can also induce hypothermia, as can various medications and toxins, including ethanol, particularly as related to alcoholism. Episodic hypothermia is associated with lesions in certain areas of the brain and is often associated with seizure disorders.[5] There are thus many conditions that make us unable, or less able, to regulate our temperature, and in mainstream culture we typically categorize some of them as being more psychological than physiological.

## The psychological consequences of impaired thermoregulation

We can go on. Thermoregulation seems to be responsible for, or at the very least involved in, psychological functioning more generally. Hypothermia resulting from prolonged exposure to the cold has sometimes been associated with psychiatric symptoms, including anxiety, impaired judgment, perseveration, neurosis, and psychosis. Recall also that the Social Thermoregulation and Risk Avoidance Questionnaire (STRAQ-1) my colleagues and I created revealed that people who want to regulate temperature on their own tend to be more anxious, while people scoring low on a desire to regulate temperature with others are more avoidant in their relationships. And those who tend to desire social thermoregulation less also report being in somewhat poorer health. In a 1966 report to a medical commission on accident prevention, L. G. C. E. Pugh reported various cases from Wales, Scotland, and England involving the inability to regulate temperature. Twenty-three exposure incidents were reported on, involving 25 deaths, five cases of people becoming unconscious with recovery, and 58 milder cases. In the milder cases, Pugh reported that patients became apathetic, sometimes preceded by anxiety. Some became more detached from reality, to the point that they told Pugh they felt inebriated. Their partners reported the patients to be irrational, irritable, aggressive, and unusually silent.[6]

Also notable is what the authors of the article "Accidental Hypothermia" call the lack of "appropriate adaptive behavior." For instance, "paradoxical undressing" was widely reported among those suffering from hypothermia. "Clothes are removed in a preterminal effort to address thermoregulatory collapse."[7] Alcoholic inebriation increases the risk of paradoxical undressing, which could result in dangerous behaviors. (Something to keep in mind the next time your friend has had a few too many in the throes of a winter pub crawl.)

Whatever the dire physical consequences caused by significant deviations from our optimal body temperature, it is likely that how we feel can have a considerable effect on our body-temperature regulation. A 2001 study sought to identify the mechanisms and mediators of rise in core temperature caused by specifically psychological stress. The authors explained that, although numerous case reports concerning "psychogenic fever" exist, it is unclear how (or even if) psychological stress raises core body temperature. In my group's work on the relation between diversity of social networks and core body temperature, we asked study participants to report on their own stress levels. The resulting dataset revealed that the effects of stress on core body temperature were about the same as the level of humidity outside, whether someone had sugary drinks, or whether someone felt anxious to be without their smartphone. In other words, stress did not greatly affect core body temperature in our large dataset from 12 diverse countries.[8]

The finding that stress has a minimal effect on core body temperature differs from the results of some studies on psychological stress–induced rise in nonhuman animals' core temperature. For instance, Takakazu Oka and Tetsuro Hori, two Japanese researchers in psychosomatic medicine and integrative physiology, reviewed studies of "open-field stress" in rats that were removed from their cages. The researchers concluded that there is a rise in core temperature after stress. In these studies, the increase in core temperature seems not to be the result of increased physical activity (as in the case of shivering, for example) but is more likely due to an elevated thermoregulatory "set point," as often happens in fever.

Oka and Hori also review studies on increases in mice's core body temperature caused by what they call anticipatory anxiety stress. The researchers believe this kind of stress is different from open-field stress, as it is inhibited by different neurotransmitter substances. The anticipatory anxiety stress also produces a fever-like response, which can be experimentally altered. Whether induced by open-field stress or anticipatory anxiety stress, the researchers concluded that the resulting temperature rise is a fever. Moreover, because the fever response can be conditioned, Oka and Hori speculate that mechanisms involving neurotransmitter substances might be involved in human psychogenic fever.[9]

There are considerable complexities in the relationship between stress and core body temperature as observed in rodents. Yet these animals' social lives are much less complex than those of humans. I suspect that this compounded complexity is why we don't really find much that translates to human stress as related to core body temperature. What we find is just so much statistical noise.

## Depression and anorexia nervosa

The famous clinical psychologist Aaron Beck once emailed me his observation that his clinically depressed clients often reported feeling cold.[10] This would likely surprise no one. Feeling cold and feeling depressed seem "naturally" to go together. Yet this tells us nothing about the physiological mechanism that makes the link seem so obvious.

A 1970 study by Anthony Wakeling and Gerald Russell investigated the regulation of body temperature in 11 female patients with anorexia nervosa, a potentially life-threatening eating disorder characterized by self-imposed food restriction, an overpowering desire to be thin, and an intense fear of gaining weight. (Eleven healthy female patients were also included in the study as a control group.)[11] Although most anorexics are underweight, they see themselves as too fat. As little as they eat, some induce vomiting or abuse laxatives.

We know that heart damage, osteoporosis, infertility, and other disorders associated with malnourishment may develop as a result.

The 11 patients were tested in their hospitalized, malnourished state and were subsequently retested after eating. The tests consisted of measuring oral and skin temperature in response to a heat stimulus and a standard meal. Researchers found that these patients essentially became less sensitive to any kind of change in temperature. Anorexia nervosa thus seems to be linked with an impaired ability to thermoregulate. The clinical evidence is consistent with disturbances in body-temperature control. Patients exhibit cold, blue extremities; suffer from tissue and capillary bed damage; and often complain of feeling cold. Indeed, some who become severely malnourished contract hypothermia that may even prove fatal.

It is often difficult to study these anorexia populations, as we cannot reach large samples easily. Small sample sizes mean that, to date, our conclusions are mostly speculative. But our theoretical conclusions are consistent with the idea that food intake is regulated by structures in the hypothalamus, the site of temperature regulation. Surgically destroying the nucleus (called the ventromedial nucleus) within the hypothalamus leads to overeating and obesity in rats,[12] whereas bilateral lesions in the extreme lateral portions (remote from the nucleus) of the rat hypothalamus result in self-starvation and subsequent death.[13] New studies have extended these results to other species. For example, studies in both goats[14] and rats[15] are consistent in suggesting that the ventromedial nucleus is involved in regulating food intake. Note that the hypothalamus is very small in the human brain and has thus been notoriously difficult to study. Nevertheless, in humans, observations have shown that disease of this brain structure may bring about either obesity or emaciation.[16]

While we have seen that portions of the hypothalamus function as a kind of master thermostat in body-temperature regulation, we have also noted that the structure is no mere thermostat. It also figures in regulating various basic metabolic processes, sleep, fatigue, circadian rhythms, and attachment behaviors. Again, we researchers must be on guard against our susceptibility to reverse inference. The hypothalamus is complex, and one neural region is not responsible for just one behavior or mechanism. We know that its preoptic-anterior region is the site of the heat-dissipation control when the body is under thermal stress.[17] The link between food intake and thermoregulation is not random. Both are metabolic and are involved in the indirect control of the body's energy balance. Remember that Wakeling and Russell studied anorexia patients, and in their work they speculated that disorders of food-intake regulation can be related to the inability to regulate body temperature.

Early theories of the causes of anorexia nervosa emphasized psychological origins, including the emotional trauma of childhood sexual abuse and growing up in dysfunctional families. Other psychological factors believed to contribute to anorexia nervosa have included anxiety, loneliness, low self-esteem, and depression. Sociological causes, based on cultural body ideals, have in the past also figured prominently. More recently, studies have examined genetic factors (the disorder is highly heritable) as well as an overactive hypothalamic–pituitary–adrenal axis (leading to an inability to regulate hormones well). The connection between anorexia and depression is not simply the causal link some have suggested. As with anorexia, early theories of depression portrayed it primarily as a psychological disorder, whereas more recent work investigates disorders of the body and social context. In this, both medicine and psychology continue their long but steady voyage toward treating neural functioning as bodily phenomena, with the brain, nervous system, and every other bodily component encompassed within a single biological whole.

While theories locating the brain as the source of all mental illness have certainly not been swept aside, there is increasing recognition that mood disorders, especially depression, are probably more accurately characterized as brain-body disorders involving both the central nervous system and the peripheral nervous system as well as all that impinges on the central nervous system. This reflects a still-developing view of psychological health not just as grounded in the brain but as a larger, more inclusive system that adapts to the physical and social environment; that is, input from the body to the central nervous system plays a key role in both cognitive and emotional states. Among the input from the periphery are thermosensory signals, which are likely to play an important role in both a sense of well-being and depression.[18]

Traditionally, theories have focused on the physiological aspects of thermoregulation in achieving and maintaining homeostasis. Recent studies, however, have furnished evidence suggesting that the neural mechanisms involved in regulating body temperature are linked far more intimately with emotional states than traditional theories have acknowledged. We have seen that exposure to physical warmth and cold are associated with cognitive and emotional behavior relating to social warmth and cold. There is a body of recent studies (in rodents) suggesting that physical warmth stimulates the production of serotonin. In popular culture, these neurotransmitters are associated with producing feelings of well-being, happiness, and even euphoria. To an extent, this view holds some truth, though the biochemical and physiological realities are considerably more complex. In any case, the preclinical rodent studies suggest that, by activating the serotonin-producing neurons, physical warmth produces something like an antidepressant effect. We can thus conclude that thermosensory pathways interact with the brain systems controlling emotions, and that not being able to regulate temperature well may be associated with affective disorders. Most intriguing is research suggesting that applying physical warmth—that is, activating warm thermosensory neural pathways—may have therapeutic potential for treating affective disorders, including depression.

We know that people with affective disorders exhibit altered perception of temperature and altered responses to changes in skin temperature, while they are not always able to regulate body temperature. Some researchers even think that skin-conductance level may be a marker for depression.[19] As with many of the purported links between mood and temperature, based on the available evidence I don't agree on the existence of simple biomarkers for psychological syndromes. With or without discrete biomarkers, depressed individuals seem to have a problem regulating their body temperatures. In line with Beck's suspicions voiced to me via email, depressed patients do show altered responses to heat.[20]

Negative emotional responses to normally innocuous thermal stimuli may also be associated with depression, since it reduces perception of pleasant warmth while increasing perception of unpleasant heat.[21] Studies have also shown that people with depression sweat less than healthy people, indicating a poorly functioning cooling mechanism. A 2009 meta-analytical study of results from three independent laboratories involving 279 depressed and 59 healthy subjects suggests that reduced sweating resulting from lower electrical skin conduction is likely a marker for suicide risk in depressed patients.[22] Studies show that incoming thermosensory signals stimulate serotonin-synthesizing systems and brain areas associated with depression.[23] This implies that the body's cooling mechanisms don't work well in depressed people.

A probable link among social factors, thermoregulation, and depressive disorders is suggested by a 2007 study examining the relationship between atypical depression and self-comforting behaviors (such as a craving for a "comfort" food like chocolate and a desire to warm up with a hot bath). Are these behaviors used to counteract low skin temperature or social coldness? Might they be an effort to trigger cooling mechanisms that tune down both sympathetic and emotional arousal along with core body temperature? Or could both motives be present? Recall our experiment in which social exclusion led to lower skin temperatures and where holding a warm beverage reduced the negative effects of social exclusion. I believe that physical warmth can alleviate depressive feelings to some degree, but the real solution is of course more complicated. It lies in a relationship among the social environment, temperature, and how you cope with these factors. In the next few years, I believe we will finally have the technology available to study these links in detail.

# Turn down your temperature, turn up your health?

Wim Hof was born in my native Netherlands, but he's become internationally famous as the Iceman. His dual professions are extreme athlete and marketer of the Wim Hof Method (WHM), which combines forced cyclical breathing techniques and meditation with extreme cold exposure to improve physical and emotional well-being. The breathing-meditation combination is similar to Tummo, a Tibetan yogic practice named after the Tibetan Buddhist goddess of heat and passion, used by tantric yogis to gain control over bodily psychophysical energy. Hof has employed the WHM to set Guinness world records for swimming under ice (188.6 feet/57.5 meters), running the fastest barefoot half-marathon on ice and snow (2 hours, 16 minutes, 34 seconds), and full-body contact with ice (setting the world record 16 times, most recently, 1 hour, 53 minutes, 10 seconds). In 2007, Hof climbed to 23,600 feet (7,200 meters) on Mt. Everest in nothing but shorts and shoes, failing to summit because of a foot injury.[24]

These achievements are remarkable, but Hof's boldest claims are for the physical and emotional health benefits of WHM, which include improved immune system, improved mental health, enhanced sports performance, reduced stress, more energy, better sleep, faster workout recovery, increased willpower and concentration, relief from depression, recovery from burnout, relief from fibromyalgia and arthritis, amelioration of the lingering symptoms of Lyme disease, better management of asthma and COPD, enhanced creativity, and improved cold tolerance.[25] Once again, you can count me as skeptical. No intervention can make good on such incredible, not to mention non-credible, claims.

While Hof does commercially market WHM training, I don't question his motives or integrity. In fact, I regularly practice his cold-shower training myself, and people close to Hof have described him to me as a thoroughly remarkable, kind, and honest person. He has been open and very cooperative with scientists eager to study what is essentially the willful regulation of autonomic regulatory mechanisms, especially during exposure to cold. Hof believes that he has found methods for improving both physical and mental health through temperature. A 2018 study by Otto Muzik and others used fMRI and PET/CT imaging to study the effects of the WHM on the sympathetic nervous system (which primarily stimulates the body's fight-flight-or-freeze response) and the way muscle and fat tissue consume glucose. The fMRI analyses suggested that the WHM activates the brain's primary control centers for modulating pain/cold stimuli, perhaps triggering a stress-induced pain-relieving response. The WHM was also shown to have an impact on higher-order cortical areas of the brain, namely the anterior and right middle insula, which are associated with self-reflection.

It appears as if the activity from the WHM may facilitate internal focus and sustained attention in the presence of unpleasant or stressful stimuli, including cold. Significantly, however, it has not been proved that the WHM can activate BAT. As for the breathing techniques of the WHM, imaging studies found that forced breathing increased sympathetic neural activity in the muscles that run between the ribs and help move the chest wall. This generates heat that warms lung tissue, which, in turn, warms the blood circulating in the pulmonary capillaries. Most significantly, the 2018 results seemed to provide evidence for the role of the central nervous system over the body in mediating Hof's responses to cold exposure. The central nervous system is associated with voluntary neural mechanisms, whereas the peripheral nervous system is associated with autonomous mechanisms. Researchers believe that their studies suggest "the compelling possibility that the WHM might allow practitioners to develop higher level of control over key components of the autonomous system, with implications for lifestyle interventions that might ameliorate multiple clinical syndromes."[26]

An earlier study from 2017 used PET/CT scans to measure Hof's BAT activity. These researchers found that his Tummo-like breathing–meditation technique produced increased metabolic activity and an increase in nonshivering thermogenesis, which, at 40 percent, was high "but not that extreme." This was, lead researcher Wouter van Marken Lichtenbelt concluded, hardly miraculous. He believes Hof withstands such extremes as immersion in ice cubes through combining "increased heat production" and "conserving the body core heat by [using] his well-trained mental ability to endure the cold (change of mindset, as he calls it)." On the other hand, it is also reported that once Hof "steps out of the ice, he starts shivering just as everybody else."

None of this disproves Wim Hof's claims, but it does imply a need to be more modest about them. Recall my skepticism about some of the studies published in neuroscience. The studies on the WHM are no different. Both categories suffer from sample sizes that are too small for us to be confident about their conclusions. Van Marken Lichtenbelt credits studies indicating that "mild cold can have profound effects on health." Exposure to extremely cold air (39.2°F/4°C for 20 minutes a day) increases the capacity of nonshivering thermogenesis, and the WHM may increase the spontaneous process by which blood vessels constrict. Moreover, the WHM does seem to improve the mental health—or at least the *feeling* of emotional well-being—for those who practice it. Van Marken Lichtenbelt notes that "many of Wim Hof's pupils become extremely enthusiastic, experiencing their bodily feelings during almost bioenergic hyperventilation sessions that they never experienced before. They are also pushing boundaries and feeling relaxed after extreme body challenges." He concludes, nevertheless, that the effects of the WHM "on our health wait to be proven," yet "people may *feel* healthier."[27]

There is a final wrinkle in the Iceman's saga. Wim Hof has an identical twin brother. Identical twins share the same DNA, offering researchers a rare opportunity to distinguish the effects of nature from those of nurture. Wim Hof's twin leads a relatively sedentary lifestyle, in contrast to his "extreme" brother, yet shows similar BAT activity. This suggests that relatively high BAT mass and activity is a genetic feature the twins share and not primarily the result of any special activity or training. Thus, there may be a significant limit on what the practice of a regimen like WHM can achieve to "train" autonomous thermoregulatory and metabolic mechanisms. In any case, no studies exist on the WHM and subsequent social behavior, but it is clear there *should* be a link. It is unclear, however, whether the link will prove to be unreservedly positive.

## Temperature as therapy

Despite the caveats, Wim Hof's experience does suggest that temperature can be therapeutically effective. Although the 2017 study (on a very small number of subjects) found that the WHM increased BAT activity in Wim Hof himself, the results were "not extreme." A study published two years earlier also could not demonstrate that cold exposure could increase BAT mass and activity. But this study did show changes in the glucose transporter of the muscle, providing promise for diabetes therapy: 10 days of cold acclimation (57.2°F–59°F, 14°C–15°C) increased insulin sensitivity by about 43 percent in eight type 2 diabetes patients.[28]

Fascination with the Iceman has cast a spotlight on the benefits of cold temperatures. On the other hand, numerous researchers have now investigated the possible therapeutic effects of warm temperatures, using whole-body hyperthermia to treat people with depressive disorders. Sometimes used to treat Lyme disease and metastatic cancer, whole-body hyperthermia heats the entire body to temperatures between 102°F and 109°F (39°C and 43°C), and even higher, using infrared domes, hot rooms, wrapping with hot wet blankets, or having the patient wear a tubed suit through which heated water is circulated.

The mechanism by which whole-body hyperthermia works has been studied in rodents, suggesting that elevated temperatures activate serotonin-producing midbrain neurons. Both the substance and the neurons are implicated in thermoregulatory cooling as well as antidepressant and antianxiety behavioral effects.[29] Based on the rodent studies, some researchers believe that whole-body hyperthermia in humans could reduce depressive symptoms by lowering core body temperature.

Why would heat create the apparently paradoxical effect of lowering core temperature? It is possible that external heating activates and/or sensitizes the neural circuit running from biosensors in the skin and other bodily tissues to areas of the midbrain and then back to the periphery. If depressed people suffer from a dysfunctional skin-to-brain-to-skin circuitry, it is likely that their impaired thermoregulatory cooling would result in elevated core body temperature. If we increase external body temperature, thermoregulatory cooling may be stimulated and produce a beneficial antidepressant response.

So, does cold training cure or ameliorate a variety of ills? Does heating the body cure or ameliorate depression? Or is much more involved?

Researchers have studied how phase relationships among core body temperature, melatonin, and sleep are associated with depression and its severity. They came to believe that misalignment of circadian rhythm—the sleep-wake cycle that repeats roughly every 24 hours—was implicated in nonseasonal depression. Some sought to determine the degree of misalignment by measuring the time interval between dim-light melatonin onset, which indicates timing of the body's central circadian pacemaker, and midsleep. The data from such studies suggested that circadian misalignment may play a role in depression.

In 2011, a new study widened the picture to include the possible misalignment of the dim-light melatonin onset with core-body minimum temperature and midsleep. The study implied that circadian misalignment was more complex, involving more systems than previously thought. The researchers were able to replicate the earlier relationship between circadian misalignment and depression severity in patients with major depressive disorder. But they also found that a misalignment between midsleep and core-body minimum temperature was associated with greater depression. Furthermore, they discovered preliminary evidence of an association between the severity of depression symptoms and the misalignment of dim-light melatonin onset and core-body minimum temperature.[30]

Thus, depression emerged as a disorder possibly related to three misalignments: lack of circadian rhythm synchronization with midsleep, dim-light melatonin onset, and core-body minimum temperature. This insight should deepen our respect for the hypothalamus not as the body's thermostat but as its great integrator. It's remarkable that different regions of this small brain structure are associated with things as varied as body temperature, fatigue, sleep, circadian rhythms, hunger, thirst, and aspects of attachment behavior. Moreover, we know that a region within the hypothalamus regulates melatonin secretion.[31]

## Cancer, heat, and the BAT connection

As we have just seen, thermoregulation operates as part of a dynamic system of causes, effects, and feedback loops. Therapeutic interventions involving cold or heat may or may not affect this system beneficially. Only extensive experiment, trial, observation, and analysis can produce insights that may have medical applications. But our attitude toward such diseases as diabetes, depression, and—perhaps most urgently—cancer is not one of patience. The potential benefits of adding thermal therapy to the arsenal of cancer treatments is appealing.

In modern medicine, cancer and heat often go together. Hyperthermia, in which body tissue is exposed to temperatures as high as 113°F (45°C), is increasingly being employed in clinical trials. The

objective is to use heat to kill or damage cancer cells while caus-
ing minimal injury to healthy tissue. Heat attacks the cancer cells,
damaging proteins and structures within them and thereby potentially shrinking tumors. At present, hyperthermia is typically used in
conjunction with more traditional treatments, such as radiation and
chemotherapy. While local treatment techniques are used to attack
small areas, such as an individual tumor, regional hyperthermia can
be applied to entire limbs, organs, or body cavities, and whole-body
hyperthermia is being tested as a treatment for metastatic cancers
that have spread throughout the body.[32]

Heat also plays a role in cancer detection. High-resolution infrared imaging can be used to detect skin cancers early in their development,[33] and studies are underway to test whether it can be used in place
of surgical biopsy to determine whether a skin spot is malignant or
benign. Because malignant melanomas have higher metabolism and
increased blood flow, they may exhibit slightly higher temperatures
than healthy skin, and high-resolution infrared imaging appears to
be sufficiently sensitive to detect such temperature elevation.[34]

While these emerging and experimental thermal methods of
detecting and treating cancer are promising, I'm more interested in
the roles of dysfunctional metabolism and thermoregulation; indeed,
these mechanisms may promote the development and growth of can-
cerous tumors. So far, the most suggestive experiments in this field
of inquiry have been done only with mice. As we know, BAT plays
an important role in body-heat generation and metabolic balance.
Two mouse studies are especially intriguing for what they suggest
about how BAT may contribute to cachexia, the wasting syndrome
often found in cancer.

As a 2012 study observes, cancer cachexia and cancer-related
anorexia involve metabolic imbalances so severe that they are considered the proximate cause of death in 20 to 30 percent of all cancers.
Researchers investigated how specific types of bowel tumors affect
BAT in mice and disturb the ability to engage in the synthesis and
degradation of lipids in cells. In response to this, BAT in mice is acti-

vated to generate heat. Moreover, body-heat generation is present in the dark cycle of the circadian sleep-wake rhythm, which normally is associated with lower, not higher, body temperature. The researchers conclude that, in activating the BAT to generate body heat, the weight loss from this type of cancer in mice stimulates an energetically wasteful and therefore maladaptive response to anorexia. They further note that, because the tumor-bearing mice were housed at 71.6°F (22°C), a temperature that causes some cold stress in mice, the BAT activation may have been due to the compromised animals' inability to maintain core body temperature.[35]

This study is significant for its suggestion that cancer is related to the dynamic mechanism of metabolism, circadian rhythm, and thermoregulation. That is, the wasting effects of cancer are associated with metabolic and thermal dysregulation. In mice, at least, both tumor growth and BAT are activated in cachectic cancer, with lethally maladaptive consequences where the body cannibalizes itself to maintain the thermal homeostasis essential to survival. As suggestive as the mouse study is, it is a very wide leap from these small animals to humans, particularly because mice have far more BAT than humans.

A study from 2016, designed to explore the initiation and progression of human breast cancer, used adipocyte xenografts (the transplant of an organ, tissue, or cell to an individual from another species) from human breast-cancer tumor cells to mice.[36] The researchers focused on the behavior of these adipocytes, which are specialized for the storage of fat and are known to promote breast cancer development. The transplant increased the expression of BAT markers in both the mouse cells and the breast tumor cells introduced into the mouse. Among the effects the transplant produced was an increase in COX-2, a protein that is responsible for inflammation and known to stimulate formation of beige adipocytes. Treatment with a COX-2 inhibitor, SC236, reduced tumor growth.

When researchers injected mice with factors that induce BAT development in vitro (in the lab, outside the body), the mice developed larger tumors. The researchers also found that xenografts derived from established breast tumor cells as well as from patients' tumor tissues expressed BAT markers and contained cells resembling BAT adipocytes. Taking these findings together with the evidence that reduction of BAT activation shrinks tumors and that BAT activation may play a role in the development of breast tumors, the researchers speculated that BAT may be a likely experimental target for the creation of a therapeutic breast cancer drug.

The following year, 2017, saw publication of a human study further investigating the role of BAT in breast cancer. Researchers found that the presence of BAT is associated with the expression of human epidermal growth factor receptor 2 (HER2), whereas the absence of BAT may be prognostic of breast cancer. Research reported in 2018 on a sample of 142 patients with a variety of cancers (breast, lymphoma, lung, gastrointestinal, melanoma, genitourinary, thyroid, and sarcoma/carcinoma of unknown origin) and showed that BAT activity is greater in people with active cancer than in comparable BAT-positive patients who do not have an active malignancy.[37] This suggests that BAT plays a role in the progression of cancer (although the study is correlational and thus suggestive, at best). PET scans using fluorodeoxyglucose, a tracer substance that indicates metabolic activity in tissue corresponding to glucose uptake, revealed the presence of BAT in all 142 patients, but those with active cancer had higher BAT volume compared to patients without active malignancy. Thus, while BAT is often expected to be of positive influence in combating obesity, the BAT story is not all positive, and as the lead 2018 researcher concluded, "modulation of BAT may play a role in cancer therapy" in the future.

## Live together or die alone

Experimental researchers in thermoregulation have clearly reached a scientific frontier, and researchers in the medical/pharmacological community are right there with them. The prospects are exciting for directly thermal therapies—both body cooling and heating—as well as for therapeutic drugs, biologicals, and gene therapies. We do not have any studies in humans that support the relationship between BAT and attachment or social networks. The reason for this lack is technological; until recently, the only way to detect BAT was via radioactive tracers using computed tomography (CT) scans, which made BAT research too invasive for reasonably large-scale studies. At the moment, we are only guessing—though it is an educated guess—that people who are relatively disconnected, isolated, and alone may have more BAT.

The implication—which demands far more research to substantiate—is that attachment to a diverse social network distributes the metabolic burden of thermoregulation and thus reduces the need for BAT, perhaps analogously to the way in which human connection to a diverse network more efficiently meets thermoregulatory demands than does *physical* proximity to other warm bodies. If social connection does distribute the metabolic burden of thermoregulation across the group, the adequately connected individual who takes on more calories than her level of activity requires—and that includes the level of activity required for thermal homeostasis—will put on weight. The relatively isolated person, in contrast, lacking a group onto which he may distribute the metabolic burden of thermoregulation, may resort to developing and recruiting BAT to keep warm. The late John T. Cacioppo, a University of Chicago professor who cofounded the field of social neuroscience, likened the desire for social connection to an urgent biological drive, like hunger.[38]

Self-help strategies such as the WHM temperature training, medical interventions, and the therapeutic use of either low or high ambient temperature offer promise for improving physical and mental health, but successfully achieving and maintaining sustained thermal homeostasis requires abundant connection to a social network. This is borne out by a valuable meta-analysis, published in 2010 by my good friend Julianne Holt-Lunstad and her colleagues.

The researchers wanted to quantify the link between people's social relationships and the chance of dying, as well as to compare this likelihood of mortality to other, well-known benchmarks of health and longevity, such as smoking, exercising, and alcohol consumption. Julianne and her colleagues analyzed 148 studies, representing a total of 308,849 participants. They were interested in the strength of people's relationships and how this factor related to how soon the people would die. What they discovered was nothing less than stunning. The complexity of people's social relationships was the strongest predictor for how long people live and even stronger than other commonly cited risk factors, such as not drinking six glasses of alcohol per day, not being obese, not getting the flu vaccine, and not smoking up to 14 cigarettes daily. How powerful this effect is can be seen in the figure below. This certainly does not mean that your relationships will give you license to adopt one or all of these high-risk behaviors. But the finding does imply that social connection should be one of the most important foci of societal health interventions.[39]

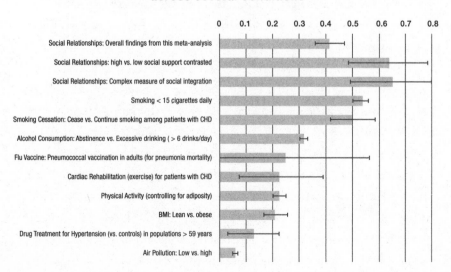

**Comparison of the chances of having lower mortality across several conditions**

Another large-scale meta-analysis by Julianne and her colleagues, published in 2015, specifically examined loneliness and social isolation as risk factors for mortality. What they found in this follow-up meta-analysis was at least as stunning as the findings from the previous study. The researchers found that a one-point increase in feelings of social isolation was associated with a 29 percent increased likelihood of mortality; a one-point increase in loneliness, with a 26 percent increase; and a one-point increase in living alone, with a 32 percent increase in the likelihood of death.[40]

## Don't buy biomarkers

Researchers often base their hypotheses on perceptions backed by collective experience, received wisdom, broadly shared assumptions, and that compelling driver of belief called common sense. They then design their experiments accordingly and too often end up with a reductive, though formally unobjectionable, conclusion. The flaw in this approach is very often the result of a data sample that is too small. Correcting the flaw, therefore, usually involves increasing the size of the sample, whether by recruiting a large number of participants or by performing a meta-analysis of data from many previous studies. Either method of compiling a large database can produce some surprising results.

Here is a case in point: SAD. Seasonal affective disorder was first reported and named in the early 1980s, but it may have been used to describe Scandinavians as early as the sixth century by a Goth scholar named Jordanes.[41] Indeed, SAD reflects in some measure the common and commonsense belief that "bad" weather, especially the cold and gloom of winter, makes many (perhaps most) people feel more depressed than "good" weather—fair, warm, and bright.

A 2010 study by researchers from the Netherlands (a country known for gloomy weather and considerable changes in seasonality) wanted to determine the causes for depressive symptoms caused by seasonal changes. The researchers reported data from a large screening program conducted in the southern Netherlands (latitude 51°15'N) to recruit participants for a depression treatment study. From a larger sample of 217,816 people, the researchers relied on the participants from the first 12 months, totaling 14,478 responders. The seasonal prevalence of major depression and sad mood (as classified in DSM-IV) in this sample was calculated and linked to the numerical mean for daily temperature, duration of sunshine, and duration of rainfall. The prevalence of major depression and sad mood did show seasonal variation, with peaks in the summer and fall.

One surprise is that there was no peak in winter, generally accepted as the prime SAD season. The more significant surprise, however, was that weather conditions were not associated with mood and therefore did not explain the seasonal variation the researchers found. From this analysis they concluded that, contrary to popular belief and common sense, weather conditions and sad mood or frank depression do not appear to be associated, not even in a country with a winter as gloomy as the Netherlands.[42]

It appears, then, that winter does not commonly or simply cause depression. In our own research, we also did not find self-reported stress to be related to core body temperature. Generally speaking, commonsense associations between weather and emotion go to the core of what we call "biomarkers." The World Health Organization has defined biomarkers as "any substance, structure, or process that can be measured in the body or its products and influence or predict the incidence of outcome or disease."[43] That means that the biomarker must be accurate and reproducible. It is not so easy for psychological syndromes. In 2014, Sami Timimi wrote that, despite a long search for reliable biomarkers, researchers have not found them for psychiatric disorders.[44] The same goes for thermoregulation and cancer or depression. The inability to thermoregulate cannot simply be seen as a biomarker for depression. Reality is much more complicated than that. Taking just depression, my colleague Eiko Fried published a paper in 2016 revealing that seven instruments to measure depression encompass at least 52 disparate symptoms.[45] It would thus be highly surprising if depression could be detected simply by measuring core body temperature.

Or, put more intuitively, it would be surprising if a researcher manipulated temperature and the participant reported back feeling positive because she was warmer. I am writing this at the end of a blistering hot summer in Grenoble (one of France's hottest cities in the summer) on a rainy day. I can report back that I am not gloomy or depressed but instead am delighted for a few drops of rain after so much heat.

Social thermoregulation's central theme is about regulating toward a temperature homeostasis. Experiences of positive or negative affect provide information on how well a person is doing to reach temperature homeostasis. Experiences of depression are more complicated than that but at least perform a similar function. Successfully regulating your temperature involves a strong social network, an ability to regulate temperature, and other variables like height, weight, and gender. Vital for our health, temperature regulation does play a crucial role in our lives. We learned in Chapter 5 that social networks are key to regulating body temperature, while social networks are more important to survival than abstaining from six glasses of alcohol per day.

In the final chapter, we will talk about social thermoregulation and place, including ideas about "national character" and well-being as it is supposedly formed and deformed by temperature and weather. This will be a plunge into a body of time-honored ideas, traditions, and received wisdom. Well-being, as we will see, is an important and timeless, yet surprisingly difficult, subject.

## CHAPTER 9

# The Happy Costa Ricans

## *Temperature, Weather, and Well-Being*

We've discussed the effect of temperature on marketing and selling real estate, but ask realtors themselves to list the three most important components of a property's value, and you will almost certainly get this knee-jerk response: "Location, location, location." This cliché was reputedly coined in the 1920s by British real estate tycoon Lord Harold Samuel, and it has certainly proved durable. The longevity suggests that this real estate marketing mantra says something important about human history and social behavior. Associating location with happiness is well established in the history of civilization. In 1516, the English social philosopher, statesman, and martyred Catholic saint Sir Thomas More gave the most extreme philosophical expression of this association a name, Utopia. It is the title of his book about an imaginary ideal island-state, but he didn't invent the concept; the notion of a place that promises infinite happiness is common in mythology, religion, literature, popular culture, and commerce. In fact, it drives the entire vacation industry.

Tao Yuanming (365?–427), a poet of the Chinese Six Dynasties period, wrote "The Peach Blossom Land" (or "The Peach Blossom Spring"), a lyrical description of a land of happiness hidden from the world, and the very phrase still figures in the Chinese language as the equivalent of the English *utopia*. The very first terrestrial location mentioned in the Old Testament is the Garden of Eden, no mere utopia but an earthly paradise. Eden has served as the benchmark against which people raised in the Judeo-Christian cultural tradition have been measuring the degree of happiness associated with places from Maui to Disneyland, Tahiti to Las Vegas. No less a figure than Christopher Columbus was confident he had found the Garden of Eden on his third voyage to the New World (1498–1500). From Hispaniola, the Caribbean island that is today divided between Haiti and the Dominican Republic, he wrote to his patrons, Isabella I of Castile and Ferdinand II of Aragon, "I believe that the earthly Paradise lies here . . . this land which your Highnesses have commanded me to discover." The admiral described, among other things, the land's "very mild climate."[1]

Columbus's discovery of "Eden" seemed to promise infinite happiness but ended up launching centuries of warfare between European conquerors and Native Americans, as well as among the competing conquerors themselves. Without question, the quest for "promised lands" has driven people to extremes of expense, exertion, risk, and wholesale slaughter. At least part of this intense activity over history has been associated with climate. We already know that thermoregulation is second in urgency only to breathing, and we have formed a good idea of how closely intertwined thermoregulation is with our social relationships. Furthermore, we have seen that temperature can have profound effects on us emotionally and cognitively. It can make us feel more or less lonely and more or less inclined to invest, to buy a variety of products, and especially to evaluate, buy, or refrain from buying real estate.

So, let's return to a question that people—from explorers to pioneers to monarchs with conquest on their minds—have been asking since who knows when. Does what we know about temperature and our well-being mean that places with "very mild climate" host happier and healthier populations?

# Weather and well-being

There is a body of research suggesting that weather can have an impact on our emotions and behavior. Many of us harbor an intuitive or commonsense notion that depression is strongly associated with long, cold, gray winters. Yet in Finland, suicide rates correlate with the onset of relatively high springtime temperatures. Analyses done on suicide in Finland suggest it is increased sunlight, not decreased temperatures, that correlate with an increase in suicide. Again, it is not just temperature that triggers a certain behavior (in fact, this study specifically indicates that temperature has only a minimal relationship with the number of suicides). In this case, BAT is the potential explanation, since, following the long cold winter, its activity may impair thermoregulation in suddenly warmer conditions, thereby increasing the risk of suicide.[2]

There is also research suggesting more directly that certain climates are more conducive to happiness than others. Studies exist that indicate people are the most satisfied with their lives in countries and regions where monthly mean temperatures do not deviate far from about 65°F (18°C).[3] So, is this thumbs-up for Costa Rica, Rwanda, and Colombia and thumbs-down for Russia, Finland, and Estonia? Should the Russians all flee to Costa Rica?

Not so fast.

History, mythology, and common sense have consistently associated happiness with place. Columbus was not deranged in his belief that he had found the Garden of Eden in Hispaniola. He merely drew conclusions based on a powerful shared cultural value from the Old Testament books of Genesis and Ezekiel. In our own era, large and ambitious surveys of well-being around the world affirm that the value of happiness is deeply embedded in national cultures. A prime example of this is the World Values Survey, which grew out of the European Values Study, first conducted in 1981 under the leadership of Jan Kerkhofs and Ruud de Moor of Tilburg University in the Netherlands, where I used to work. These sociologists were test-ing the hypothesis that modern economic and technological changes transform values and motivations in industrialized civilization.[4]

Since 1981, the World Values Surveys have been conducted in six "waves" over an expanding geographical and cultural sample. After the first wave in 1981, a second was conducted from 1990 to 1991, a third from 1995 to 1997, a fourth from 1999 to 2001, a fifth from 2005 to 2007, and a sixth from 2010 to 2014. In 2015, the seventh wave was launched. The World Values Survey is vast, with data cur-rently from some 400,000 respondents in nearly 100 countries that together contain almost 90 percent of the global population. It cov-ers very poor to very rich countries in all of the world's major cul-tural zones. The intention is to provide a database to aid scientists as well as policy makers in appreciating the beliefs, values, and motiva-tions of populations throughout the world. The conclusions reached thus far, the leaders of the survey argue, support the hypothesis that positive economic development, democratization, and increased tol-erance elevate people's sense that they have freedom of choice. This, they argue, leads to higher levels of happiness.

Indeed, while researchers have catalogued a list of "30 most cru-cial findings," all of them relate in some way to happiness. Thus, the leading insight of the World Values Survey may well be that happi-ness is an integral part of virtually all national cultures, although such measurements rarely feature in national policy—at least not yet.

While the World Values Survey does not specifically study correla-tions between climate and happiness, various follow-up studies have suggested that unhappy "survival" cultures evolve in poor countries that have colder-than-temperate or hotter-than-temperate climates, both of which place pretty severe demands on people's need to ther-moregulate. In contrast, these studies suggest that easygoing, mod-erately happy cultures evolve in countries with temperate climates, regardless of per capita income. The happiest cultures are those that exhibit "self-expression values"—aspiration to liberty, social toleration, life satisfaction, and freedom of public expression—in contrast to mere "survival values." "Self-expression" cultures evolved in rich countries that nevertheless have the kind of colder-than-temperate or hotter-than-temperate climates that make significant thermoregulatory demands.

I have no doubt that happiness is a significant cultural issue. This insight is famously enshrined in the United States' 1776 Declaration of Independence, whose principal author Thomas Jefferson borrowed from the Enlightenment philosopher John Locke in enumerating three "unalienable" human rights: life, liberty, and "the pursuit of happiness." It is significant that Jefferson took life and liberty directly from Locke but replaced Locke's original third right, "property," with the concept of happiness.

I am also convinced that climate is significant in shaping our happiness. But first we must define happiness in more granular terms. Let's provisionally accept that recent research correlates the moderate happiness of easygoing cultures with temperate climates irrespective of wealth or its absence, whereas unhappy survival cultures are associated with intemperate climates—but only when coupled with the presence of poverty. Change poverty to wealth, and intemperate climate becomes associated with the greatest degree of cultural happiness. Read Jefferson's words closely. He did not simply swap Locke's *property* for *happiness*. He wrote of the pursuit of happiness, connoting anything but a passive state. *Pursuit* suggests exertion, the investment of effort and energy, including the metabolic energy required to thermoregulate in a climate that presents temperatures significantly below or above the thermoneutral zone.

In exploring the relationship among temperature, weather, and well-being, we once again find ourselves tempted to jump to easy conclusions about temperature and emotion and, therefore, temperature and behavior. The research reported here does show the impact weather may have on our lives, but I will explain why these effects are neither as strong nor as directly simple as common sense and popular belief suggest. Hopefully these insights will help us make better educated guesses about how to interpret temperature fluctuations and their effects on our happiness.

Temperature is an aspect of environment that oftentimes requires our conscious attention and effort. Some environments are easier to cope with than others. Those that vary least from the thermoneutral zone require very little metabolic energy. But more extreme thermal conditions challenge us. Some of us are not up to the challenge, but others will invent such technologies as central heating and central air-conditioning. Technology demands a high degree of network diversity and social collaboration, which may be the reason that our thermoregulation becomes involved even in the maintenance of more complex social networks. In poorly organized societies or ones short on certain crucial natural resources, temperature extremes may bring great misery, sickness, and death. In more developed societies, however, especially those with access to strategic resources, the challenges posed by intemperate climate accelerate technology, promote commerce, and build wealth.

As we move further into the twenty-first century, however, weather extremes will pose fresh challenges as the global climate undergoes unprecedented changes, created by the cultural influences of evolution. Feeling cold and sad? Reaching for a warm cup holding hot tea will probably make you feel better, at least for a short bit. But it will not improve society's ability to cope with climate change. Culture is not the individual writ large, a mistaken assumption that scientists as well as civilians frequently make. On the individual, community, national, and global levels, we must learn to cope effectively with temperature and thermoregulation. Doing so is vital to our happiness, health, and longevity—as well as our more immediate survival. This starts with acquiring an understanding of social thermoregulation so detailed that together we can devise the social means of coping with new challenges to our thermal homeostasis. But temperature does not make us happy or sad, rich or poor. It challenges us to cope, to adapt, to invent. Governments must foster an understanding of our human penguin nature by ambitious research programs so thorough that they yield knowledge on which scarce revenues and other resources can be more effectively allocated to design and execute socially based coping programs. We cannot all move to Costa Rica for the climate.

## It's so SAD

Seasonal affective disorder, or SAD, just makes sense. If the onset of winter—the chill in the air, the falling leaves, the bare branches, the outward migration of the birds—makes you sad, then you are not alone. We might assume that SAD is the poster-child disorder "proving" the notion that weather, especially temperature, makes us unhappy, even depressed. The National Institute of Mental Health (NIMH), a U.S. government agency, defines SAD as "a type of depression that comes and goes with the seasons, typically starting in the late fall and early winter and going away during the spring and summer. Depressive episodes linked to the summer can occur but are much less common than winter episodes of SAD." The NIMH lists a set of symptoms, including a subset of the symptoms of major depression, and stipulates that a diagnosis of SAD requires meeting "full criteria for major depression coinciding with specific seasons (appearing in the winter or summer months) for at least 2 years."[5]

But recall from Chapter 8 the study where the authors concluded that weather conditions do not seem to be associated with a sad mood or depression. Speaking from personal experience, I find it especially compelling that the authors of this study are associated with the University of Maastricht and the University Hospital Maastricht. If researchers working in a country with weather as fickle and gloomy as that in my native Netherlands cannot find strong evidence for SAD, just where else is such evidence to be found?

No animal species is more widely distributed on the planet than people, a fact that testifies to our incredible ability for adapting to our environment, including new and even extreme situations that may be posed by weather. Recall from Chapter 3 my conclusion that endothermy allows humans to be far more flexible to living in a wide variety of environments. If you think about thermoregulation and depression from this perspective, a broad relationship between weather and mood actually seems unlikely. Given humanity's endothermic flexibility, we should expect that humans have the potential to adapt to new environments with new ambient temperatures, and that even the individual's very personality changes to cope with the demands of a new environment or the extremes of a familiar one.

A 2017 study by Wenqi Wei and others tested just that. It sought to uncover the relationship between ambient temperature and personality using data compiled from 59 Chinese cities and a dataset from 12,499 U.S. zip-code locations. The researchers defined as "clement" temperatures at or near 71.6°F (22°C) and found that people who grew up in places with more clement temperatures scored higher on personality factors associated with socialization and stability, such as agreeableness, conscientiousness, and emotional stability, as well as factors correlated with personal growth and "plasticity," such as extraversion and openness to new experience; this was compared to individuals raised in regions with less clement temperatures, who scored relatively lower in these factors. (Note that the researchers included in their analyses only those individuals who had grown up in the areas under study.) Yet Wei and colleagues also found that individuals in less clement environments are able to adapt, on the level of personality, to the demands of the climate.

Now, the finding concerning personality admittedly complicates the relationship between temperature and mood. Apparently, there can be a robust relationship between climate and personality, suggesting that humans adapt but that these effects on personality might be activated only if they are long term. I must admit that in the beginning I was skeptical of the studies by Wei and team. Journal editors usually ask scientists to do quality-control checks of others' work before a publishing decision is made. The reviewers to whom the editors reach out typically have special content expertise. In full disclosure, I was tapped as a reviewer for the Wei article and recommended rejection rather than publication. I am glad the editor overruled me. I did, however, request additional analyses. The authors complied—and proved that I was wrong (again!).

My doubts notwithstanding, the work of Wei and colleagues did make me think about the relationship between their study and our Human Penguin Project (HPP), which we discussed in Chapter 5. Among many other things, we found that people who live farther away from the equator score higher on social diversity than those living closer to it. Locations farther from the equator are likely associated with less clement temperatures, at least on the colder end of the thermometer. Does this tend to contradict the finding of Wei and others that individuals in less clement environments score lower on such factors as agreeableness, openness to experience, and extraversion than those in more clement locations? Not necessarily. You can conclude from Wei and colleagues that people in more clement climates are more positive. But the more interesting conclusion is that people in less clement climates readily adapt to their environments.

My takeaway from the Wei study is that those in less clement conditions are highly adaptive to their demanding weather and climate. Do you necessarily have to be a "nice" person to participate in a variety of relationships, from very close and loving ones to those based more on reciprocity, as with strangers, or on financial exchanges, as with your bank manager? Being from the Netherlands, where European capitalism was essentially invented, I must say no. We know that the extremely inclement conditions of the penguins' Antarctic environment drive their social thermoregulatory huddling. Penguin huddling requires a large group, but modern human beings create social thermoregulation with network diversity rather than network size. The diversity of these groups results in more complicated relationships, not all of which require people to be "nice," let alone very agreeable. Thus, insofar as network diversity is a social adaptation to less clement temperature conditions, the Wei study seems to be consistent with our findings in the HPP.

We humans do not simply or inevitably surrender to temperature extremes. Our evolutionary nature to adapt implies that any attempt to meaningfully correlate temperature with emotion and behavior will be complicated by our stubborn refusal to become soft targets and passive victims of the thermometer. Technological and other social adaptations make relationships between temperature and behavior, as well as temperature and personality, both difficult and fascinating to study. Moreover, as part of the HPP, we developed the STRAQ-1 to measure individual differences in the desire to socially thermoregulate, and we found that individuals do indeed significantly differ in this. Whatever else this tells us, it is a reminder that while culture is not the individual writ large, neither is the individual the culture writ small—nor, for that matter, is any individual a universal representation of the entire species.

If we extend social behavior to include sexual behavior, it may be useful to observe that many insect species become increasingly polyandrous—meaning that females mate with multiple males— as temperatures rise, but others remate more frequently as temperatures go down. In the case of the fruit fly species *Drosophila pseudoobscura*, a species extensively used in laboratory studies, the increase in polyandry is associated with cooler temperatures. The female of this species remates more frequently in geographic locations at higher latitudes—that is, farther from the equator.[6] While there is an unfortunate dearth of data on the relation between temperature and human sexual behavior, it is true that having sex can, through both exercise and skin-to-skin contact, be a way of dealing with falling temperatures. But whether polyandry provides a way for women to deal with the retreating mercury in colder climates is yet unknown. We do know that Evelyn Satinoff concluded that human sexual and thermoregulatory behaviors do indeed overlap—not, however, because of any causal relationship but because the brain has evolved similar mechanisms for various forms of motivated behavior. Both thermoregulation and sex are strongly motivated by the same areas in the brain, even if they are otherwise mostly unrelated to each other.

What, if anything, does adaptation have to do with SAD? Statisticians and scientists distinguish main effects from interaction effects created by other independent variables. For example, a student performs poorly on exams. Studying more and harder is a likely main effect that would improve performance. So is tutoring. Reducing hunger as a distraction during test taking might have some influence on performance, but it is not likely to be a main effect. Symptoms consistent with a diagnosis of SAD do exist, but SAD is not likely to be a main effect in creating major depression. Most people adapt to the seasonal stresses, including the metabolic demands created by, for instance, lower winter temperatures and possible dysregulation of the circadian cycle as a result of seasonal photoperiod variation—for example, reduced sunlight hours in winter. Those who cannot adequately adapt to these stresses may indeed suffer from depleted energy and the emotional symptoms that accompany this state.

At the close of the last century, researchers from the University Hospital Groningen in the Netherlands published a study of the prevalence of SAD in relation to latitude. The major hypothesis described was that "SAD is triggered by photoperiod variation," which should be consistent with an increased incidence of SAD the farther we travel from the equator and lose sunlight hours. A review and analysis of 22 studies available at the time revealed that the mean prevalence of SAD was twice as high in North America as in Europe. While a significant positive correlation between SAD prevalence and latitude was found in North America, the correlation in Europe amounted to only a "trend." Although scientists often use trends as an indication that they have identified an effect, noting a trend really means that nothing of significance was found.[7]

In 1989, Leora N. Rosen and others set out to study the prevalence of SAD at four latitudes in the United States. From north to south, these were Nashua, New Hampshire; New York, New York; Montgomery County, Maryland; and Sarasota, Florida. The Seasonal Pattern Assessment Questionnaire (SPAQ) was mailed to people in these areas. Questions in this survey ranged from "At what time of the year do you feel best, do you socialize most, or do you lose most weight?" to "When do you take the most naps throughout the year?" The sample in this study was selected randomly from telephone books (remember those?) but was balanced for sex. Ultimately, 1,671 respondents returned completed questionnaires. (As the researchers themselves concede, this methodology may have created a so-called oversampling bias in that those who responded to the questionnaires may have had more of a special interest in seasonal problems than people farther away from the equator.)

The study concluded that in the more northern latitudes, rates of winter SAD and S-SAD were "significantly higher." (S-SAD is subsyndromal SAD, a condition popularly called "winter blues," in which individuals report fewer and milder symptoms than those required for a formal diagnosis of SAD.) There was no correlation, however, between latitude and prevalence of summer SAD. Interestingly, the winter SAD correlation was found mostly in people over 35 years old. The researchers noted that their study did not make clear the extent to which age, as a factor in SAD, is related to latitude as opposed to such location variables as employment opportunities, cost of living, and availability of resources for retired persons.[8] All of these factors likely influence individual resilience and adaptability.

From these results, we can conclude that latitude has a "small" influence on the prevalence of SAD but that such factors as climate, genetic vulnerability, and "social-cultural context" are more important influences on its prevalence. With varied sunlight hours (relative to latitude) discounted in the study as a leading cause of SAD, the three remaining candidates include one geographical factor—climate—and two factors that relate to the person: genetics and social-cultural context. This is consistent with the imperative of active adaptation to the demands of place and climate. But while we know a number of main effects (network diversity, individual differences in the desire to socially thermoregulate) we still know next to nothing about how they interact. It stands to reason, however, that if you live in a colder climate, you would be best off seeking out a more diverse social network.

## WEIRD world versus real world

Since at least as early as the late 1980s, SAD has been the subject of considerable study and even more conversation within the general public. Numerous studies, including those we have cited here, suggest links between SAD and temperature, sunlight, seasonality, latitude, and other aspects of weather and/or climate; however, they also point out serious shortcomings in the studies, ranging from small sample size to independent variables inadequately or entirely unaccounted for. Indeed, as we saw, the largest study, from Marcus Huibers and colleagues, asks a simple question, "Does the weather make us sad?" and concludes that no relation whatsoever between weather and depression can be established based on existing data.

This leaves us with three possible explanations. First, weather/climate is not an important variable in the incidence of depression, and SAD, as a collection of symptoms possibly associated with weather/climate, is not a main effect in creating major depression. Second, weather can make us sad, depressed, or experience SAD, but we have yet to devise studies capable of proving this. (After all, I remain a big fan of the mantra "Absence of evidence is not evidence of absence.") Third, we don't know if the first or second explanation is correct because measurement of depression is more difficult than we ever thought, and even scientists do not agree on just what symptoms constitute depression.

Once we determine the list of symptoms to study, we can address the problem of measurement. It is understood in the sciences that if you cannot meaningfully measure a thing, you cannot meaningfully study it. In the social sciences, most of the measures we have developed are focused on "WEIRD" populations, typically local university students. By this, social scientists mean young people from Western, Educated, Industrialized, Rich, and Democratic nations and cultures, typically studied by researchers in universities, as students are most convenient to study.[9] The "Big Five" personality traits—the qualities of personality that psychologists universally consider testable and measurable, namely openness, conscientiousness, extraversion, agreeableness, and neuroticism—do not work for persons in lower- and middle-income countries. In those nations, psychometric tests based on the Big Five as conceptualized for WEIRD nations and cultures fail to measure the intended traits and are therefore less valid.[10] Measurements designed with young WEIRD people in mind also prove unsatisfactory when applied to older people. The relationships among indicators within each given trait as well as across traits do not remain constant over the human life span. Psychometric tests, questionnaires, and studies must be designed with our subjects in mind.[11] As my colleagues Eiko Fried and Jess Flake would say, "measurement schmeasurement."[12]

Measuring aspects of behavior and personality is not the same as shopping for tube socks. One size does not fit all. We must agree on what to measure and how to measure it, and then we must adapt our observational techniques to the population, or populations, under study. I believe we psychologists do reasonably well measuring what we intend to measure. But our discipline has not yet reached the level of maturity of physics or chemistry when it comes to our measuring instruments and statistical models. Why not? Well, as you have repeatedly encountered in this book, studying even something as "simple" as social thermoregulation in humans consists of understanding body size, personality, culture, and how much "brown fat" you have.

Consider something central to SAD: assessing people's mood states. The study of major depression typifies psychology's measurement problem. A paper published in 2016 by Eiko Fried, a fellow fan of *gezellig* coffee houses like Native, notes that the severity of depression is assessed differently across research disciplines. There are currently seven scales commonly used to assess the severity of depression. The scales differ widely in content and, taken together, encompass 52 depression symptoms. For one thing, the volume and variety of symptoms suggests that depression has become a catchall disorder, vaguely and even idiosyncratically defined. It is therefore quite likely that research results obtained using one scale might not be replicable on other scales, let alone in other populations! This greatly complicates depression research.

Fried undertook a content analysis to assess symptom overlap across the seven scales and found that the mean overlap was very low, ranging from 0.27 to 0.40. Indeed, 40 percent of the 52 disparate symptoms appear in just a single scale, and only 12 percent feature across all seven scales. Idiosyncratic symptoms appear in some scales (at rates ranging from 0 to 33 percent), and compound symptoms appear at rates ranging from 22 to 90 percent. Fried makes no attempt to distill the 52 symptoms into a smaller number of more clearly and accurately defined symptoms, but instead ventures that the estimate of 52 imperfectly identified symptoms is likely a conservative underestimate. That is, the current state of differences across the seven different scales is probably even greater than Fried's estimate. The problem this creates is that the combination of differences across scales and minimal overlap (indicating no consensus) tends to create research results that are idiosyncratic to the particular scale or scales used, making it extremely difficult to replicate results or to meaningfully generalize from them.[13]

The severity of depression is measured based on the number of symptoms used to establish threshold scores and classify a person as depressed or not depressed. This would be a valid approach if depression really were a single condition with universally agreed-upon symptoms. But the implication of Fried's analysis is that the quantity and variety of symptoms attributed to "depression" indicate the absence of agreement not only on which symptoms define the disorder but which ones are the most valid measures of its severity. This does not at all mean that depression is neither a valid label nor a valid diagnosis. Depression is, in fact, both of these things. But it does mean that the treatment of the disorder is more complicated than we often like to believe.

Fried and another researcher, Randolph M. Nesse, cite "a host of studies" that reveal such depressive symptoms as sad mood, insomnia, difficulty concentrating, and thinking about suicide as not only distinct phenomena in themselves but phenomena that differ from one another in very significant ways, ranging from their underlying biology to the risks they pose. But lumping disparate "symptoms" into a sum-score to estimate the severity of the depression impedes research, including efforts to identify biomarkers of depression in order to find, formulate, or apply more effective antidepressant treatments.[14]

Until well into the nineteenth century, science was generally called "natural philosophy." We modern scientists generally insist on distinguishing ourselves from philosophers of any kind, but we would do well to recognize, whatever our scientific field, that we owe an intellectual debt to at least one major area of philosophical concern: epistemology. This branch of philosophy is devoted to the nature of knowledge and belief, and it warns us that we cannot take for granted the validity of what we conclude based on our observation of phenomena. In researching depression, for example, to what extent are we accurately describing and measuring phenomena, and to what extent are we assessing the language we use to identify and describe the symptoms of what we call depression? Based on the complexity involved, we must conclude that too few of us are sufficiently diligent in addressing the epistemological issues underlying our research. Are we fitting hypotheses to phenomena or merely to language?

This is the greater question Fried and colleagues ask in their critical examination of depression research. Moreover, the high rate of failure to replicate the results of so much scientific research is related to a failure to answer this question. Popular media treats the replication crisis as a crisis of ethics in academia, an epidemic of fraud. Unfortunately, while it is true that some experiments have been fudged, Fried and others have pointed out that failure to replicate results is perhaps more often due to a failure to define what should be measured and how to measure it. The epistemological objective is to get as close to the phenomena as possible without getting lost in subjective terms that are either poorly understood or ambiguously defined. This is generally an important issue, but nowhere is it more evident than in hypotheses concerning the relation among temperature, weather, place, emotions, and other aspects of well-being.

## From passivity to adaptation

We know that thermoregulation relies on the primal systems networked throughout our bodies, is coordinated by successively higher neural systems such as the hypothalamus, and ascends to the highest levels of the cortex. It also reaches beyond our bodies and brains, to other persons, social networks, society at large, and, indeed, civilization itself, including the built environment of technology. Little wonder, then, that we encounter one finding that is consistent above all others: our world is far more complex than can be comprehended in hypotheses based on the assertion of simple relationships.

Let's think again about the complexity of our social world and how an individual's behavior cannot be easily extrapolated to culture. Scientists sometimes abridge the features of complexity. One example is a paper by Paul van Lange and others published in 2017, which attempts to explain "aggression and violence around the world" with a model reducible to a conveniently catchy acronym, CLASH: CLimate, Aggression, and Self-control in Humans. The hypothesis is that the world presents a picture of aggression and violence that shows "substantial differences."[15]

You can certainly entertain various hypotheses as to the causes of greater aggression, one of them being climate. This is not necessarily a good explanation, as evidenced by commentaries on the target article by Van Lange and his colleagues, but let's run with the hypothesis. It has been suggested that people in warmer climates simply interact more frequently: the routine-activity theory. But there can be alternative theories. There are, for example, various experimental studies showing the effects of warmer temperature on violence.[16] The typical explanation is that warmer temperature may make people lose control and lash out. The short story behind the CLASH article is that, based solely on these effects, the authors felt justified in extending the behavior of the individual to the culture, thus hypothesizing that living in warmer climates makes you more aggressive because you have lower self-control. Their reasoning is this: lower temperatures and a greater range of seasonal variation in climate (both of which are found farther from the equator) prompt individuals as well as groups to focus more on the future as opposed to the present. (Recall the Warren Buffett metaphor we adopted earlier, indicative of a slow life history.) Therefore, the authors propose the hypothesis that people living in countries farther away from the equator have better self-control.

I was skeptical of this hypothesis when I first encountered it. My colleagues and I were conducting the Human Penguin Project right around the same time, and Van Lange and his colleagues themselves suggested that data-driven methods were needed to investigate their hypothesis. Just having picked up machine learning for the HPP, we could devise a perfect test. What's more, our dataset should be able to detect what they predict even more easily. In our datasets the effects should be amplified, as the "data points" were somewhat far away from each other: we had only one country really close to the equator, meaning the climate effects should be exaggerated, which would thus favor the Van Lange hypothesis. In our analysis of the CLASH project using the HPP, we collected latitude, self-control, and many social predictors from a dozen countries at varying distances from the equator and analyzed the data from 1,507 participants.[17]

We found a very weak effect between the distance from the equator and self-control. But with such large datasets, you are bound to find significant effects, which are mainly due to overfitting, discussed in Chapter 5. The effects are not real. More meaningful is the comparison of distance from the equator to other variables. How important is distance from the equator in determining the level of self-control? Just about as important as speaking Serbian—in other words, not important at all.

I am a champion of more-complex stories of human development, like Jared Diamond's *Guns, Germs, and Steel*.[18] The CLASH model ignores many of the factors we discussed earlier in this chapter— namely, the great range of influences, effects, and variables that impinge on individual human behavior and on that of regional and national populations. Notably, we did find another important predictor of self-control: whether people are anxious in their attachment. In other words, social environment, as opposed to climate, predicted self-control—and therefore, potentially, aggression. Human culture and psychology are thus not the simple sum of individuals, and therefore they resist reduction to the sweeping generalization and oversimplification of one model.

Recall also that intemperate climates seem to be associated with unhappiness only when people are unable to deal effectively with the demands of those climates. The reason why I am a fan of Diamond's work is that he points to very complicated historical developments, including the fact that wealth is often accumulated through mere historical accidents: it was "simpler" for people in Europe and Asia to exchange information because traveling east to west was far easier than across climates from north to south in the Americas; European armies carried germs to the Americas, which nearly wiped out entire (and much larger) opposing forces; and Europeans encountered higher availability of animals that could be domesticated, therefore allowing for far greater food supplies. To better understand how well we can control ourselves, climate must be examined together with the diversity of our social networks, our wealth, how we individually cope with climate, and, finally, the luck of the draw.

The reality of complexity should not, of course, discourage attempts at good science. Earlier in this chapter, I discussed Wenqi Wei's paper on ambient temperature and human personality. Let's take a second look at the paper in the broader context of its ambitious hypothesis. The authors do not argue that geography somehow creates regionally characteristic personality traits but instead present the hypothesis that humans adapt to their environment and that this human capacity to adapt extends even to something as apparently fixed and unchangeable as personality. Wei and colleagues thought that because all humans continually experience and react to ambient temperature, it stands to reason that temperature is a "crucial environmental factor . . . associated with individuals' habitual behavioural patterns." This being the case, temperature must also affect even what we know as the most basic aspects of our personality.

Recall that the researchers collected data from a large Chinese sample and an even larger U.S. sample. Accepting the definition of personality as an interactive aggregation of characteristics influencing individual response to the environment, the authors argue that the so-called Big Five personality traits can be gathered into two comprehensive factors, which they designate Alpha (agreeableness, conscientiousness, and emotional stability) and Beta (extraversion and openness to experience). The Alpha traits are associated with socialization and stability; the Beta traits with personal growth and plasticity.

Reasoning that humans have an existential need for thermal comfort, the researchers proposed that mild temperatures are conducive to exploration beyond immediate shelter. This enables a greater range of social interaction and experience. In contrast, extreme temperatures in either direction make people less likely to venture outside unless doing so is absolutely necessary. In consequence, those who live in very hot or very cold climates socialize less and try fewer new activities. The authors thus predicted that people who come of age in more mild temperatures score higher on the Alpha socialization factor and the Beta personal-growth factor. Using their large cross-cultural datasets to support the association of regional ambient temperature with personality traits, they were able to explain why personalities vary across geographical regions in ways that cannot be fully explained by such older theories as subsistence style theory, selective migration theory, and pathogen prevalence theory. Looking forward in the context of global climate change, the researchers predicted that significant temperature-related changes in personality may emerge in the foreseeable future.

The breadth and depth of the study's samples won me over to the paper. But equally persuasive was what the researchers concluded about those populations in less mild environments. They were hardly doomed to living out their lives as solitary troglodytes. The study's findings were consistent with what we know about social thermoregulation. As we have learned, people in cold environments do not wither like feckless houseplants. Driven by social thermoregulation wired through the higher cognitive centers of the brain and influenced by a cultural evolution rooted in genetic evolution, people seek "social warmth" in the creation of diverse social networks. Even the Dutch, dwelling in their gloomy climate, seek out the *gezelligheid* offered in the likes of Native and Brûlerie des Alpes. Analogous establishments find popularity in most cultures. Rather than surrender and succumb to the thermal environment in which they happen to find themselves, people build a cultural, social, and technological environment with their given one far more effectively than a network of huddling penguins.

The temperature clemency hypothesis—people who find themselves living in mild climates get out more, enjoying the open-air marketplaces and meeting places, the sun-drenched agoras of classical Greece—does not contradict the social thermoregulation hypothesis. In both clement and inclement climates, people seek society. Where the weather is pleasant, they find it easy to mingle. It's a party on the balmy beach! But where the weather is more challenging, posing threats to the precious and precarious window that is thermal homeostasis, people also seek each other, perhaps even more urgently and with greater ambition. In both environments, people evolve society.

There is an even more important similarity that transcends place and climate. Even if clement and less clement temperatures foster the development of different personality traits, neither condition "makes" people behave one way or another. Instead, both enable or motivate people to adapt to their particular world. The effect of temperature on humans is not the same as the effect of temperature on, say, yeasty bread dough. Dough rises in a warm environment; it has no choice in the matter. The action appears dynamic but is, in fact, entirely without volition. In contrast, people respond to their environment with varying degrees of dynamic adaptation. The success or failure of the adaptation we choose and activate speaks volumes about the relative health or dysfunction of each of us.

# *Afterword*

René Descartes was key to the Scientific Revolution. Although born a Frenchman, he had the great good sense to live 20 of his most productive years in the Dutch Republic. I come neither to praise nor to bury him, but I have written a book that argues vigorously against his highly influential dualist vision of mind and body as proximate but separate, like the "pilot" who guides the "ship" that carries him but of which he is not a part.

What compelled me to argue against the Cartesian vision? Well, simply everything I have learned in my research on thermoregulation in general and social thermoregulation in particular. Most compellingly, separating the mind from the body assumes that we are very different from other animals. Alas, we are not.

Not everyone agrees with this. As we discussed at length in Chapter 2 and elsewhere, Lakoff and Johnson's conceptual metaphor theory posits that abstract concepts are represented in concrete experiences because we coexperience them. They believe we learn the symbolic concept of affection through experiences of physical warmth by coexperiencing both the affection concept and physical warmth when we are held by our caregivers. This, Lakoff and Johnson argue, is an obvious example of embodied cognition: a symbolic, and thus cognitive, concept that comes to be understood through bodily experience. For them, embodied cognition is the sovereign corrective to Cartesian mind-body dualism.

And yet it is no such thing. For while the concept and its physical correlative (and vice versa) are experienced together, the resulting metaphor is first and last a product of the mind. The pilot is still in command of the ship, if you will. This necessarily implies that social thermoregulation is something we acquire as the mind develops. But even if we assume that the acquisition takes place in early infancy, a bird's-eye view of findings from behavioral ecology, physiology, and developmental psychology does not support social thermoregulation as an acquired mechanism. Rather, it is something we are born with. Moreover, it is something we share with other homeothermic endotherms. Making their homes in Antarctica, penguins live as social animals, constructing their social behaviors not according to some conceptual metaphor (what do penguins know of metaphor?) but as a way of staying warm enough to survive. The individual neurological and physiological mechanisms that drive and enable them to organize their mass huddles create their complex social behavior. It is social thermoregulation, with no conceptual metaphors required.

Penguins cannot create metaphor. People can. Yet when it comes to social thermoregulation, people and penguins are one and the same. Neither species relies on metaphor to enable the social behavior so crucial to thermoregulation. Lakoff and Johnson do not acknowledge our penguin nature. Just because we humans can create metaphors—some of them quite wonderful—does not mean that all of our experiences of the world are mediated by them. Some experiences, including social thermoregulation, are enabled without conceptual metaphor. The ability to create metaphor sets us apart from other animals, but the life-giving necessity of thermoregulation, including social thermoregulation, directly connects us to them.

✦  ✦  ✦

Let's return to Sheldon Cooper of *The Big Bang Theory*. In Chapter 1, notice that Sheldon does not just tell us that offering a hot beverage to someone who is feeling sad or disappointed will raise their spirits. Indeed, he frames the offer as "a nonoptional social convention." Even though Sheldon is a fictional character and not actually a scientist, his nonoptional social convention takes us not just from the Cartesian brain-body dualism to brain-body continuity but, from this, to a continuity among brain, body, and society.

Thermoregulation reconciles the divorce of mind from body by yielding a profound insight into what it means to be human, while research into social thermoregulation extends the reconciliation beyond the individual. So, while penguins huddle and humans weave diverse networks, the evolutionary root and impetus for these strikingly different behaviors are the same in both species. The body of evidence points to the role of thermoregulation in how we form and maintain interpersonal relationships, true in humans as well as in other animals. People are penguins—but with culture. (Among other things, this should remind us that temperature is not the only force driving social behavior. Yet again, we must not make the reverse inference, however tempting, that cold temperatures make an individual more capable of being social.)

I have not been shy about pointing to how we psychologists have often been wrong. When I speak of the "body of evidence," I don't deny that not every study supports this body. I myself have been wrong at various times. In my defense, I'll point out that being wrong presents the opportunity to correct erroneous predictions, which in turn presents the opportunity to learn. But the evidence for the role of social thermoregulation in creating relationships is ample, and it gives me the confidence to overcome my own hesitation when it comes to applying research findings. I believe and predict that we can use the insights of social thermoregulation to modernize relationship therapy in a manner that colleagues and I call Social Thermoregulation Therapy. STT can enhance existing relationship therapies by integrating sensor and actuator technologies into the well-established approaches of Emotionally Focused Therapy to help people use thermoregulation to improve the quality of their close relationships.

The stakes in this endeavor are high. A large body of work demonstrates that successful relationships are among the very strongest predictors of physical and mental health as well as happiness. Such relationships are predictive of your "life chances," including longevity, enhanced creativity, and higher self-esteem. To date, research on how relationship quality influences life chances has focused on what we may call the "higher-order" levels, defined in couples experiencing fewer marital problems, having generally better health, and enjoying greater satisfaction with their relationship.

My colleagues and I turned from such higher-level results to lower-order thermoregulatory issues: health problems caused by dysregulated body temperature, the role of temperature regulation in sociality, the reliance of social warmth on physical warmth, and a specific lower-level dynamic we discussed in Chapter 5 and elsewhere, called coregulation. Recall that *coregulation* describes an individual's continuous action or behavior as modified by his or her partner's continuously changing actions or behavior. With respect to STT, my colleagues and I asked whether thermoregulation is crucial to physiological coregulation in the close relationships of couples. We believe the answer is yes. And that elicits a second question: Can therapies be developed to improve physiological coregulation in couples? At the moment, we are trying to find out in our research whether couples coregulate each other through their peripheral and core body temperatures, and we are poised to find out whether applying STT to complement existing therapeutic approaches can help people tune up or strengthen their social lives.

In their 1992 study, John M. Gottman and Robert W. Levenson showed that coregulation is necessary to the relationship's success;

they found that positive exchanges with a marriage partner correlate with a lower chance of divorce and better general health. Years later, Emily A. Butler and Ashley K. Randall reasoned that partners' perceptions, actions, and even physiology are influenced by some continuously interactive connection between them, the actual mechanism of which remains obscure. We have already seen that ostriches can spend more time eating, with their heads low to the ground, if other ostriches in their group keep their heads up and on the lookout for predators.

Homeothermic animals off-load the metabolic demands of thermoregulation by huddling or by simply living in groups, like the *Octodon degus*. In humans, this off-loading behavior has driven the development of diverse networks and technological innovation. Think of it this way: when you intuit that your partner is angry, you can help down-regulate that anger by discussing an intellectually demanding or complicated subject matter. Or if your partner becomes sad, you can give a hug. We suspect that temperature changes are foundational to emotions in our relationships. But how does this work exactly? What is the mechanism? We are not sure yet.

Emotionally Focused Therapy (EFT) focuses on helping couples optimize coregulatory patterns in their interactions. Sometimes one partner responds to the other's anger in a way that is not conducive to feeling safe for either of them. EFT helps partners identify those aggravating patterns and improve well-being. We suspect that, by assessing how couples' temperature changes as a function of their peripheral and core body temperatures, we can begin to understand the interactions underlying the destructive patterns of behavior and expression. Thus, a chunk of why people behave the way they do in their relationships can be studied by focusing on temperature regulation. This makes it possible to use STT to complement EFT by adjusting a relationship's peripheral temperatures in ways that enhance the individuals' perception of social predictability within the relationship.

Advances in electronic health are creating digital wearables, such as bracelets with temperature sensors, as a means of assessing the thermoregulatory dynamics of coregulation in a particular relationship. In order to know whether social thermoregulation couples therapy can work, we need to study temperature-regulation patterns in couples who perceive their relationship to be of high versus low quality—and, perhaps, are even in therapy if they deem it to be low quality. If we are able to distinguish via temperature the patterns present in high-quality relationships from those in low-quality relationships, we think that we can also improve relationships via temperature. By using a smart algorithm to connect sensors with digital devices that manipulate skin-level temperature, could it be possible to make responsive temperature adjustments in couples to push low-quality relationships toward high-quality ones? Overall, we hope that doing this will beneficially change couples' behavior, even if just a little bit. Ideally, this manipulation will enhance both partners' feelings of safety and predictability in the relationship. It may move them to act within their relationship less like the day-trading Wolf of Wall Street and more like the long-term-investing Warren Buffett.

At present, the therapeutic application of STT to EFT using sensors and actuators is in an early study stage, but it shows strong potential. (Note that changing temperature will not change an abusive relationship into a high-functioning, high-quality relationship.) As of this writing, we are already using the Embr Wave Bracelet, commercially available from Embr Labs, which allows users to adjust their skin-level temperature by electronically cooling or warming one spot on the body to improve overall comfort and mood. This is a significant step toward integrating wearables into the ever-growing Internet of Things, technology that has revolutionized temperature regulation in homes and businesses.

Pause, take a breath, and imagine the implications. For most of us, the modern digital era is one that has extensively interconnected us through devices. Yet it seems more and more that people report feeling disconnected. Some surveys indicate that, in the United States, the fraction of people reporting that they feel lonely seems to have risen from 11 percent to 26 percent. For people who are 45 and over, the percentage is even higher, around 40 percent. If, however, what I have presented in this book is at least 10 percent accurate, the proliferation of social thermoregulation wearables will remarkably extend current communications technologies by enabling us to send a bit of warmth along with our voice and video image. Of course, we will need to determine when this remote temperature-manipulation technology is appropriate—and when it is not. Skyping or FaceTiming a dose of warmth may work with your partner-for-happily-ever-after or perhaps the one for a night. It may work less well when you are engaged in an online job interview with strangers.

Admittedly, my optimism about an STT-EFT future might seem a bit out of character. After all, I have repeatedly expressed my reservations and doubts about the state of psychological science. But in all honesty, I believe there are steps we are taking to make our psychological insights far more practically useful. The replication crisis, which emerged after 2011, has revealed that many of our findings are insufficiently precise and insufficiently generalizable. Yet we do know that there are some general patterns that we find time and again. To me, this implies that we are capable of identifying general phenomena but that we don't yet have the capacity to apply psychological science with a high degree of accuracy outside of the laboratory, in the "real world."

This is because humans are, you guessed it, extremely complicated. The general principles of social thermoregulation are true and real, but how, in what situations, and with whom we can apply sensors to measure and actuators to manipulate is something that requires more study. I have great confidence that we will become proficient in this application within the next 5 to 10 years. If I write a new edition of this book in that time frame, I am confident that I will report to you how therapies founded on the principles of social thermoregulation work, and work effectively.

Here's the reason for my optimism. First, psychologists have led the charge to make our science more reliable by creating standards for replication and data sharing. The biggest driving force behind this has been the Center for Open Science, led by Brian Nosek. In 2017, I was one of the cofounders of the Psychological Science Accelerator (PSA), a network of over 500 laboratories in more than 60 countries (and counting), which is developing new standards to study humans—and doing so by the day. I like to believe that, if properly funded by the relevant agencies, the PSA will form psychology's equivalent of CERN in Geneva or the UK Biobank, a resource that will allow us to study humans across different countries and with considerable sample sizes. The PSA's very first study, for example, involved over 10,000 participants. Thus I have considerable optimism that psychology will attain the level of precision seen in physics, such that researchers can report data and analysis in a common, universal, and neutral language. Without this capability, psychology cannot be either applied or evaluated meaningfully.

I am excited by this prospect, which promises not only to bring a significant measure of resolution to the replication crisis but, in fact, to introduce into psychology an unprecedented level of rigor, accuracy, and mutual understanding. It will be nothing less than a revolution, and it will enable the application of truly revolutionary social technologies—such as STT sensors/actuators—that will help us to employ, beneficially, the insights of social thermoregulation. Digital technology not only will accelerate those insights but will also affect how we interact with one another. I hope to talk to you again in 10 years, armed with all these new insights about our social world.

✦   ✦   ✦

We psychologists are getting better at applying our insights. Our own analyses in social thermoregulation studies are showing this. It's a very good thing for this field, for we must get much better—and do so quickly. Circumstances are becoming increasingly urgent.

Even as we remain animals, our one-on-one relationships lead seamlessly to the creation of networks, to society, culture, and technology. The most intimate of scales broaden instantly to assume a global scope. As a planetary species, we are in the midst of global climate change, a problem we have created. On the hopeful side, people adapt to climate in the most remarkable ways. The earth is truly our house, and we have shown ourselves capable of living in every room of it. This is the confidence I have often heard voiced: technology can resolve the problems technology has created. But there are limits. We cannot always keep up. If one room begins to catch fire, we'd better find a way to put that fire out—directly and expeditiously. If we are forced to leave that room because we can no longer stand the heat, we'd better appreciate the likelihood that the fire will spread and burn the whole house down.

Evolution takes time, whereas climate change, measured against the evolutionary time scale, is very, very sudden. Genetic evolution has positioned us to create cultural evolution and, with it, science and technology—the very things that have created an environment of accelerating climate change. Beyond the problems this causes in our physical world, climate change will also affect how we deal with one another, as is evidenced by the research in this book. Fortunately, though science and technology have played a major role in creating a climate crisis, they can also accelerate our ability to adapt, not only to take steps to ameliorate some of climate change's worst effects, but to adapt ourselves to what we cannot alter or alter rapidly. We must now take conscious and deliberate actions to build a socially rational perspective that enables us to change with climate change—to amplify, enhance, and accelerate our evolutionary faculty of adaptability.

As for this book, my hope is that it has already advanced our collective adaptability, however slightly, by providing new insights, first into the vast aspirations and second into the hard limitations of scientific inquiry. The arrogant cannot adapt. Adaptability requires humility.

Beyond this insight, I do hope I have succeeded in introducing as many interested people as possible to the powerful, central, and pervasive roles that regulation, thermoregulation, and social thermoregulation play in our individual, intimate, and collective lives. Social thermoregulation is essential to survival, and it is just as essential to thriving. It is a concept, a phenomenon, and a mechanism that reveals us to ourselves as organisms and as humans. Social thermoregulation is a lens through which we individuals appear as what we truly and finally are: beings who need each other and who have transformed this need into neighborhoods and nations, societies and civilizations.

# Acknowledgments

No student studies alone. At the risk of leaving out many influential people (which I undoubtedly will), I will try to recollect a few. Crucial to the start of my career in social psychology was Wilco van Dijk, who was the coordinator of my research master's program at Vrije Universiteit, Amsterdam, beginning in 2004. He has been influential in the start of many careers in science, including my own. In 2006, I worked with Dov Cohen at the University of Illinois in Urbana-Champaign, and with his prodding, I moved out of the mind, the psychologist's traditional domain; after doing so, I became increasingly fascinated by the role of the body in our perceptions and behaviors. Dov is one of the most creative researchers I have met, and he has been a great inspiration to me throughout the years.

During my PhD work at Utrecht University, Thomas Schubert frequently provided a lot of extraordinarily helpful advice; he was instrumental in my becoming a better scholar than I would otherwise be. Developing as a scientist is as much about disputation as it is agreement and accord. I have fundamental differences with George Lakoff and Mark Johnson, coauthors of the influential *Metaphors We Live By*, but it was their provocative work that first gave me the push to think about the role of the body in thinking. They also shoved me toward many other authors. I enjoyed reading the eminent mind-body philosopher Maxine Sheets-Johnstone, learned a lot about evolution from Linnda Caporael, and have been stimulated and inspired by anthropologist Alan Fiske's Online Relational Models Lab. That brings me to the European Social Cognition Network, which has always felt like a safe haven in which to present my research and learn from others.

The game changer for my research came with a visit to Jim Coan at the University of Virginia in 2014. Before that encounter, I thought about social thermoregulation from a cognitive perspective: how we form *thoughts* about others. Jim taught me to think about social thermoregulation from the perspective of behavioral ecology, in that it is not just about concepts in our mind but, with even greater priority, about outsourcing metabolic demands. The result? The world started to make more sense to me—slowly.

While I was in Virginia, Brian Nosek kindly invited me to work once each week at the Center for Open Science. Game changer number two. Science needed a revolution, and Brian and his team were starting one. I am eternally grateful for his admirable patience with the field of psychology (and with me). He hoped and expected both the science (and the researcher) to self-improve, and he has been working diligently on this for many years.

A treasured memory is my interactions with Harry Reis, Julianne Holt-Lunstad, Jim Coan, and Spike Lee during a fellowship in 2015 in Wassenaar, the Netherlands, at a period when my scientific life was not at a high point. Add to these people Sigi Lindenberg, who has been an incredible mentor and friend throughout the past five years of my social thermoregulation research. He has played a considerable role in the development of this work and, even more, has been an unfailing source of warmth.

In the years since Wassenaar, I have benefited much from conversation and collaboration with others. My science partners include (among many more) Mattie Tops and Ivan Ropovik, who shared with me their visions from different perspectives.

I am very, very glad that Université Grenoble Alpes gave me a safe scientific home, where I look forward to practicing solid science and studying social thermoregulation for many years to come. This scientific home is created by the laboratory chair, Dominique Muller, and the people I have collaborated with and am still collaborating with, including Lison Neyroud, Olivier Dujols, Elisa Sarda, Patrick Forscher, Alessandro Sparacio, Rick Klein, and Adeyemi Adetula, in no order of preference. I am grateful to all of them for their efforts, insights, patience, and input. I am glad for the university as a home, despite the fact that Grenoble is blistering hot in the summer, especially for a large animal from the Netherlands. But an enduring theme of social thermoregulation is the capacity for human beings to adapt to (almost) every climate on the planet. I will endure.

I would also like to thank Alan Axelrod, who has really coached me throughout this project, who has taught me some new insights about social thermoregulation, and thanks to whom I learned many new things about how to explain the concepts I thought I understood but now understand better thanks to his help. Further, I would like to thank my Norton editor, Quynh Do, who provided many wise and helpful comments on how to present the concepts laid out in this book and who keeps me from endlessly repeating myself.

I am also very grateful for the support of my friends throughout all these years. Without mentioning them all by name, thank you all. The same goes for my parents and my brother, who have been there throughout the worst and the best of times. This book, however, would not have been possible at all without my greatest source of warmth of all, the person to whom I dedicate its entirety, my wife, Daniela.

Evolution teaches me that social thermoregulation is one of the main reasons people feel compelled to form relationships. And what in life is more critically important than relationships?

# *Notes*

## CHAPTER 1
## Hot Beverages, Electric Blankets, and Loneliness

1. In truth, there are fundamental differences in the body's thermore-ceptors for warmth and heat. Thermoreceptors can be divided into low- and high-threshold receptors. Low-threshold receptors become activated for relatively comfortable temperatures (between 59°F and 113°F [15°C and 45°C]), while high-threshold receptors generally become activated outside of that range. While warmth is often asso-ciated with comfort, heat may be associated with noxious stimuli. Provided they are not scalding, hot beverages often give us the desir-able sensation of overall warmth—a comforting feeling—not the possibly stressful sensation of being hot.

2. Solomon E. Asch, "Forming Impressions of Personality," *Journal of Abnormal and Social Psychology* 41, no. 3 (1946): 258.

3. Lawrence E. Williams and John A. Bargh, "Experiencing Physical Warmth Promotes Interpersonal Warmth," *Science* 322, no. 5901 (2008): 606–607.

4. Arthur Aron, Elaine N. Aron, and Danny Smollan, "Inclusion of Other in the Self Scale and the Structure of Interpersonal Closeness," *Journal of Personality and Social Psychology* 63, no. 4 (1992): 596.

5. Hans IJzerman and Gün R. Semin, "The Thermometer of Social

Relations: Mapping Social Proximity on Temperature," *Psychological Science* 20, no. 10 (2009): 1214–20.

6.  The quote was attributed to Einstein by Lincoln Barnett in a series of essays entitled *The Universe and Dr. Einstein*.

7.  Kipling D. Williams and Blair Jarvis, "Cyberball: A Program for Use in Research on Interpersonal Ostracism and Acceptance," *Behavior Research Methods* 38, no. 1 (2006): 174–80.

8.  Chen-Bo Zhong and Geoffrey J. Leonardelli, "Cold and Lonely: Does Social Exclusion Literally Feel Cold?," *Psychological Science* 19, no. 9 (2008): 838–42.

9.  A. Szymkow et al., "Warmer Hearts, Warmer Rooms: How Positive Communal Traits Increase Estimates of Ambient Temperature," *Social Psychology* 44, no. 2 (2013): 167–76. Psychologists, shortly after this study, became quite strongly motivated to do a cleanup of their work, and this study was included in a large-scale "replication" effort called "Many Labs 3" led by Charlie Ebersole from the University of Virginia. To me the jury is still out on how reliable the effect is. Charlie and his team seemed to be unable to replicate the effect. When we examined their data, we discovered that the labs in which they tried to replicate our effect were much hotter than the original ones. From everything you will learn in this book, it seems improbable that the priming effect would emerge in hot conditions. When we reexamined the effect, we were able to replicate, but *only* under lower temperature conditions. There are still two points to critique our reanalysis on: first, we did not specify this to the replicators a priori (which is needed for prediction in science), and second, the sample size was really too small to detect this interaction (which is why the techniques we used in our reanalysis were not entirely appropriate, formally speaking).

10. George Lakoff and Mark Johnson, *Philosophy in the Flesh* (New York: Basic Books, 1999), vol. 4.

11. Hans IJzerman et al., "Cold-Blooded Loneliness: Social Exclusion Leads to Lower Skin Temperatures," *Acta Psychologica* 140, no. 3 (2012): 283–88.

12. K. Uvnäs-Moberg et al., "The Antinociceptive Effect of Non-noxious Sensory Stimulation Is Mediated Partly through Oxytocinergic

Mechanisms," *Acta Physiologica Scandinavica* 149, no. 2 (1993): 199–204.

13. Yoshiyuki Kasahara et al., "Impaired Thermoregulatory Ability of Oxytocin-Deficient Mice during Cold-Exposure," *Bioscience, Biotechnology, and Biochemistry* 71, no. 12 (2007): 3122–26.

14. Molly J. Crockett, "The Neurochemistry of Fairness: Clarifying the Link between Serotonin and Prosocial Behavior," *Annals of the New York Academy of Sciences* 1167, no. 1 (2009): 76–86.

15. M. W. Hale et al., "Evidence for In Vivo Thermosensitivity of Serotonergic Neurons in the Rat Dorsal Raphe Nucleus and Raphe Pallidus Nucleus Implicated in Thermoregulatory Cooling," *Experimental Neurology* 227, no. 2 (2011): 264–78.

16. E. Satinoff, "Neural Organization and Evolution of Thermal Regulation in Mammals," *Science* 201, no. 4350 (1978): 16–22.

17. Helen Shen, "The Hard Science of Oxytocin," *Nature* 522, no. 7557 (2015): 410.

18. William Glaberson, "After the Arguments: Jogger Jury Weighs a Jumble of Details," Reporter's Notebook, *New York Times*, August 10, 1990.

19. Christine Gockel, Peter M. Kolb, and Lioba Werth, "Murder or Not? Cold Temperature Makes Criminals Appear to Be Cold-Blooded and Warm Temperature to Be Hot-Headed," *PloS One* 9, no. 4 (2014): e96231.

20. J. Steinmetz and T. Mussweiler, "Breaking the Ice: How Physical Warmth Shapes Social Comparison Consequences," *Journal of Experimental Social Psychology* 47, no. 5 (2011): 1025–28.

21. John Bowlby, *Attachment and Loss* (London: Hogarth Press and the Institute of Psycho-Analysis, 1969); Mary D. Ainsworth, "Patterns of Attachment Behavior Shown by the Infant in Interaction with His Mother," *Merrill-Palmer Quarterly of Behavior and Development* 10, no. 1 (1964): 51–58.

22. Hans IJzerman et al., "Caring for Sharing: How Attachment Styles Modulate Communal Cues of Physical Warmth," *Social Psychology* 44, no. 2: 160–66.

23. Mary D. Ainsworth, "Infant–Mother Attachment," *American Psychologist* 34, no. 10 (1979): 932.

24. K. Bystrova et al., "Skin-to-Skin Contact May Reduce Negative Consequences of 'the Stress of Being Born': A Study on Temperature in Newborn Infants, Subjected to Different Ward Routines in St. Petersburg," *Acta Paediatrica* 92, no. 3 (2003): 320–26.

25. Jean M. Mandler, "How to Build a Baby: II. Conceptual Primitives," *Psychological Review* 99, no. 4 (1992): 587.

26. Daniel Roche, *Le peuple de Paris: Essai sur la culture populaire au XVIIIe siècle* (Paris: Fayard, 2014).

27. Great Britain Parliament, House of Commons, *Reports from Committees: 1857–58*, vol. 9: "Irremovable Poor; County Rates (Ireland); Destitution (Gweedore and Cloughaneely)."

## CHAPTER 2
## The Human Machine

1. René Descartes, *Les passions de l'âme* (Paris: Flammarion, 2017).

2. The metaphor is found in Descartes, *Meditations on First Philosophy*, "Meditation VI: Concerning the Existence of Material Things, and the Real Distinction between Mind and Body," paragraph 13. Modern readers may be confused by the use of the word *pilot* here. In the original Latin text of 1641, Descartes used the word *nauta* ("sailor, seaman, or mariner"), but in the first French translation of the *Meditations*, which was made under his supervision by the Duke of Luynes in 1647, the word *pilote* was used: *"un pilote en son navire."* Most English translations use *pilot*, including the authoritative 1901 version by John Veitch, who wrote "a pilot in a vessel." In the seventeenth century, both the French word *pilote* and the English word *pilot* were used as synonyms for "steersman" or "helmsman," the person who directs the course of a ship. This is a more specific meaning than the Latin *nauta*, but it was introduced in the first French translation, which Descartes supervised, and, apparently because of respect for that supervision, it has been generally used in subsequent English translations. For tests of all three translations, see David B. Manley and Charles S. Taylor, *Descartes' Meditations—Trilingual Edition* (Dayton, Ohio: Wright State University, 1996), https://corescholar.libraries.wright.edu/cgi/viewcontent.cgi?article=1008&context=philosophy.

3. A. M. Turing, "On Computable Numbers, with an Application to the *Entscheidungsproblem*," *Proceedings of the London Mathematical Society* s2-42, no. 1 (1937): 230–65. It is available online at https://academic.oup.com/plms/article-abstract/s2-42/1/230/1491926 and at https://www.cs.virginia.edu/~robins/Turing_Paper_1936.pdf.

4. A. M. Turing, "Computing Machinery and Intelligence," *Mind 50*, no. 236 (1950).

5. Mukul Bhalla and Dennis R. Proffitt, "Visual–Motor Recalibration in Geographical Slant Perception," *Journal of Experimental Psychology: Human Perception and Performance* 25, no. 4 (1999): 1076.

6. Jeanine K. Stefanucci and Dennis R. Proffitt, "The Roles of Altitude and Fear in the Perception of Height," *Journal of Experimental Psychology: Human Perception and Performance* 35, no. 2 (2009): 424.

7. Lera Boroditsky and Michael Ramscar, "The Roles of Body and Mind in Abstract Thought," *Psychological Science* 13, no. 2 (2002): 185–89.

8. Matthew 27:24.

9. *Macbeth* V, i.

10. Chen-Bo Zhong and Katie Liljenquist, "Washing Away Your Sins: Threatened Morality and Physical Cleansing," *Science* 313, no. 5792 (2006): 1451–52.

11. Jennifer V. Fayard et al., "Is Cleanliness Next to Godliness? Dispelling Old Wives' Tales: Failure to Replicate Zhong and Liljenquist (2006)," *Journal of Articles in Support of the Null Hypothesis* 6, no. 2 (2009); B. D. Earp et al., "Out, Damned Spot: Can the 'Macbeth Effect' Be Replicated?," *Basic and Applied Social Psychology*, 36, no. 1 (2014): 91–98.

12. Peter Brian Medawar, *The Limits of Science* (Oxford: Oxford University Press, 1984), 51.

13. Edward L. Thorndike, *Educational Psychology*, vol. 2, *The Psychology of Learning* (1913).

14. John B. Watson, "Psychology as the Behaviorist Views It," *Psychological Review* 20, no. 2 (1913): 158.

15. B. F. Skinner, *Verbal Behavior* (New Jersey: Prentice-Hall, 1957).

16. Kenneth J. W. Craik, *The Nature of Explanation* (Cambridge: Cambridge University Press, 1943).

17. Jay W. Forrester, "Counterintuitive Behavior of Social Systems," *Technological Forecasting and Social Change* 3 (1971): 1–22.

18. Even fMRI cannot provide us with definitive answers. Much fMRI research relies on too few participants. As a result, a lot of fMRI research to date is insufficiently granular to tell us very much. Studies using larger samples should yield more useful information.

19. John R. Searle, "Minds, Brains, and Programs," *Behavioral and Brain Sciences* 3, no. 3 (1980): 417–24.

20. Stevan Harnad, "The Symbol Grounding Problem," *Physica D: Nonlinear Phenomena* 42, no. 1–3 (1990): 335–46.

21. William James, "What Is an Emotion?" *Mind* 16 (1884): 188–205.

22. Robert B. Zajonc and Hazel Markus, "Affect and Cognition: The Hard Interface," in *Emotions, Cognition, and Behavior*, ed. Carroll E. Izard, Jerome Kagan, and Robert B. Zajonc (1984), 73–102.

23. One specific paradigm has been seen as crucial support for the idea that smiling leads to greater happiness or joy. This "pen-in-the-mouth" study led by German psychologist Fritz Strack had participants hold a pen in their mouth so that their "smiling muscle" (zygomaticus major) was activated without relation to smiling. After doing this (and in contrast to a condition in which the same muscle was inhibited), participants thought a cartoon from Gary Larson's *The Far Side* was funnier. A more recent large-scale replication led by Dutch psychologist Eric-Jan Wagenmakers did not find the same results. (If you are interested in how psychologists discuss these matters with one another, I recommend that you follow these discussions on Facebook. Some exchanges are rather entertaining.) Nevertheless, another meta-analysis by U.S. researcher Nicholas Coles and colleagues addressing the same idea—known as "facial feedback"—did confirm that the general concept of facial feedback is indeed operative.

24. George Lakoff and Mark Johnson, *Metaphors We Live By* (Chicago: University of Chicago Press, 2008).

25. "A Red, Red Rose," by Robert Burns:

> *O my Luve is like a red, red rose*
> *That's newly sprung in June;*
> *O my Luve is like the melody*
> *That's sweetly played in tune.*

*So fair art thou, my bonnie lass,*
*So deep in luve am I;*
*And I will luve thee still, my dear,*
*Till a' the seas gang dry.*

*Till a' the seas gang dry, my dear,*
*And the rocks melt wi' the sun;*
*I will love thee still, my dear,*
*While the sands o' life shall run.*

*And fare thee weel, my only luve!*
*And fare thee weel awhile!*
*And I will come again, my luve,*
*Though it were ten thousand mile.*

## CHAPTER 3
## Harry the Penguin

1. Aaron Waters, François Blanchette, and Arnold D. Kim, "Modeling Huddling Penguins," *PLoS One* 7, no. 11 (2012): e50277.
2. Caroline Gilbert et al., "Huddling Behavior in Emperor Penguins: Dynamics of Huddling," *Physiology and Behavior* 88, no. 4–5 (2006): 479–88.
3. Yvon Le Maho, Philippe Delclitte, and Joseph Chatonnet, "Thermoregulation in Fasting Emperor Penguins under Natural Conditions," *American Journal of Physiology—Legacy Content* 231, no. 3 (1976): 913–22.
4. S. D. McCole et al., "Energy Expenditure during Bicycling," *Journal of Applied Physiology* 68, no. 2 (1990): 748–53.
5. Bernd Heinrich, *The Hot-Blooded Insects: Strategies and Mechanisms of Thermoregulation* (Springer Science and Business Media, 2013).
6. Wouter D. van Marken Lichtenbelt, Jacob T. Vogel, and Renate A. Wesselingh, "Energetic Consequences of Field Body Temperatures in the Green Iguana," *Ecology* 78, no. 1 (1997): 297–307.
7. Natalie J. Briscoe et al., "Tree-Hugging Koalas Demonstrate a Novel

Thermoregulatory Mechanism for Arboreal Mammals," *Biology Letters* 10, no. 6 (2014): 20140235.

8.  P. J. Young, "Hibernating Patterns of Free-Ranging Columbian Ground Squirrels," *Oecologia* 83, no. 4 (1990): 504–11.

9.  A. Fedyk, "Social Thermoregulation in Apodemus Flavicollis (Melchior, 1834)," *Acta Theriologica* 16, no. 16 (1971): 221–29.

10. Adapted and simplified from "Table 3. Metabolic Savints (%) Due to Huddling in Mammals and Birds," in Caroline Gilbert et al., "One for All and All for One: The Energetic Benefits of Huddling in Endotherms," *Biological Reviews* 85 (2010): 560–61.

11. Luis A. Ebensperger, "A Review of the Evolutionary Causes of Rodent Group-Living," *Acta Theriologica* 46, no. 2 (2001): 115–44.

12. Julia Lehmann, Bonaventura Majolo, and Richard McFarland, "The Effects of Social Network Position on the Survival of Wild Barbary Macaques, Macaca sylvanus," *Behavioral Ecology* 27, no. 1 (2015): 20–28.

13. Richard McFarland et al., "Thermal Consequences of Increased Pelt Loft Infer an Additional Utilitarian Function for Grooming," *American Journal of Primatology* 78, no. 4 (2016): 456–61.

14. Robin I. M. Dunbar, "Functional Significance of Social Grooming in Primates," *Folia Primatologica* 57, no. 3 (1991): 121–31.

15. Shlomo Yahav and Rochelle Buffenstein, "Huddling Behavior Facilitates Homeothermy in the Naked Mole Rat (Heterocephalus glaber)," *Physiological Zoology* 64, no. 3 (1991): 871–84.

16. Daniel T. Blumstein and Kenneth B. Armitage, "Cooperative Breeding in Marmots," *Oikos* (1999): 369–82.

17. Jeffrey R. Alberts, "Huddling by Rat Pups: Group Behavioral Mechanisms of Temperature Regulation and Energy Conservation," *Journal of Comparative and Physiological Psychology* 92, no. 2 (1978): 231.

## CHAPTER 4
## People Are Penguins, Too

1.  Douglas G. D. Russell, William J. L. Sladen, and David G. Ainley, "Dr. George Murray Levick (1876–1956): Unpublished Notes on the

Sexual Habits of the Adélie Penguin," *Polar Record* 48, no. 4 (2012): 387–93.

2. Internet Movie Database, "Encounters at the End of the World (2007)," IMDb, https://www.imdb.com/title/tt1093824/.

3. Jonathan Miller, "March of the Conservatives: Penguin Film as Political Fodder," *New York Times* (September 13, 2005), https://www.nytimes.com/2005/09/13/science/march-of-the-conservatives-penguin-film-as-political-fodder.html.

4. Esa Hohtola, "Shivering Thermogenesis in Birds and Mammals," paper presented at the 12th International Hibernation Symposium, "Life in the Cold: Evolution, Mechanisms, Adaptation, and Application," Institute of Arctic Biology, 2004.

5. John Ruben, "The Evolution of Endothermy in Mammals and Birds: From Physiology to Fossils," *Annual Review of Physiology* 57, no. 1 (1995): 69–95.

6. James D. Hardy and Eugene F. DuBois, "Regulation of Heat Loss from the Human Body," *Proceedings of the National Academy of Sciences of the United States of America* 23, no. 12 (1937): 624.

7. An article published in the *New York Times* on August 3, 2015, on the work by Dutch scientists Boris Kingma and Wouter van Marken Lichtenbelt suggested that most buildings are based on the metabolic rates of men (and of a specific size of a man). Kingma and Van Marken Lichtenbelt pointed to the formula's critical variable, the resting metabolic rate of a 40-year-old man weighing about 154 pounds. Both in 1937, when Hardy and DuBois published their study, and in 1982, when Fanger formulated an equation for "comfort analysis" (based on 1971 data), the typical office worker probably approximated age 40, was male, and weighed 154 pounds. This is no longer the case, however. Women, who not only are smaller but also usually have slower metabolic rates than men, constitute at least half of the work force. Kingma and his colleague believe the outdated "comfort" model "may overestimate resting heat production of women by up to 35 percent."

8. Christian Cohade, Karen A. Mourtzikos, and Richard L. Wahl, "'USA-Fat': Prevalence Is Related to Ambient Outdoor Temperature—Evaluation with 18F-FDG PET/CT," *Journal of Nuclear Medicine* 44, no. 8 (2003): 1267–70.

9.    Thomas F. Hany et al., "Brown Adipose Tissue: A Factor to Consider in Symmetrical Tracer Uptake in the Neck and Upper Chest Region," *European Journal of Nuclear Medicine and Molecular Imaging* 29, no. 10 (2002): 1393–98.

10.    Unfortunately, for years, measurement of BAT was done using computed tomography (CT) scans, which rely on radioactive tracers. This technique is expensive and invasive and therefore logistically unavailable for basic research. Researchers from Australia have been developing less invasive and much cheaper measurement methods, which means we will, in the next few years, find out a great deal more about the relationship, if any, between BAT and human social behavior.

11.    Lane Beckes and James A. Coan, "Social Baseline Theory: The Role of Social Proximity in Emotion and Economy of Action," *Social and Personality Psychology Compass* 5, no. 12 (2011): 976–88.

12.    Nicholas A. Christakis and James H. Fowler, *Connected: How Your Friends' Friends' Friends Affect Everything You Feel, Think, and Do* (New York: Little, Brown, 2009), xvi.

13.    Just as thermoregulation is more complex than is accounted for by the hypothalamus-as-thermostat model, so social thermoregulation is more complex than can be explained by some recent attempts to localize the distributed thermostat in specific cortical brain structures. For instance, a 2013 paper by UCLA researchers Tristen K. Inagaki and Naomi I. Eisenberger concluded from an fMRI study that a "common neural mechanism" located in the insular cortex (a part of the outer layer of tissue of the cerebrum located within the lateral sulcus, which is the fissure that separates the temporal from the parietal and frontal lobes) "underlies physical and social warmth." As we will see in Chapter 5, this is an instance of the reverse-inference fallacy. In a 2006 article, Russell Poldrack concluded that such reverse inference (via which we link brain regions to specific cognitive processes) is not deductively valid. As researchers, we must beware of drawing conclusions via reverse inference that oversimplify the location of neural mechanisms involved in social thermoregulation. For instance, the mechanisms underlying social warmth and physical warmth are more complicated than what can be accounted for by simply locating overlap in their activation of the insular cortex as revealed by fMRI.

14. Claude Bernard, *Leçons sur les phénomènes de la vie commune aux animaux et aux végétaux* (Paris: Baillière, 1879).

15. See W. B. Cannon, *The Wisdom of the Body* (New York: W. W. Norton, 1932), 177–201.

16. Stephen W. Ranson, "Regulation of Body Temperature," *Association for Research in Nervous and Mental Disease* 20 (1939): 342–99.

17. Evelyn Satinoff, "Behavioral Thermoregulation in Response to Local Cooling of the Rat Brain," *American Journal of Physiology—Legacy Content* 206, no. 6 (1964): 1389–94.

18. H. J. Carlisle, "Heat Intake and Hypothalamic Temperature during Behavioral Temperature Regulation," *Journal of Comparative and Physiological Psychology* 61, no. 3 (1966): 388.

19. Evelyn Satinoff and Joel Rutstein, "Behavioral Thermoregulation in Rats with Anterior Hypothalamic Lesions," *Journal of Comparative and Physiological Psychology* 71, no. 1 (1970): 77.

20. Michel Cabanac, "Temperature Regulation," *Annual Review of Physiology* 37, no. 1 (1975): 415–39.

21. J. Hughlings Jackson, "On Some Implications of Dissolution of the Nervous System," *Medical Press and Circular* 2 (1882): 411–33.

22. E. Satinoff, "Neural Organization and Evolution of Thermal Regulation in Mammals," *Science* 201, no. 4350 (1978): 16–22.

## CHAPTER 5
## Rat Mamas Are Hot

1. Paul Ekman, Robert W. Levenson, and Wallace V. Friesen, "Autonomic Nervous System Activity Distinguishes among Emotions," *Science* 221, no. 4616 (1983): 1208–10.

2. Stephanos Ioannou et al., "The Autonomic Signature of Guilt in Children: A Thermal Infrared Imaging Study," *PloS One* 8, no. 11 (2013): e79440.

3. Michael Leon, Patrick G. Croskerry, and Grant K. Smith, "Thermal Control of Mother-Young Contact in Rats," *Physiology and Behavior* 21, no. 5 (1978): 793–811.

4. Leigh F. Bacher, William P. Smotherman, and Steven S. Robertson, "Effects of Warmth on Newborn Rats' Motor Activity and Oral

Responsiveness to an Artificial Nipple," *Behavioral Neuroscience* 115, no. 3 (2001): 675.

5.  Monica Nuñez-Villegas, Francisco Bozinovic, and Pablo Sabat, "Interplay between Group Size, Huddling Behavior and Basal Metabolism: An Experimental Approach in the Social Degu," *Journal of Experimental Biology* 217, no. 6 (2014): 997–1002.

6.  Harry F. Harlow, "The Nature of Love," *American Psychologist* 13, no. 12 (1958): 673.

7.  Carl Bergmann, *Über die Verhältnisse der Wärmeökonomie der Thiere zu ihrer Grösse* (1848).

8.  Richard McFarland et al., "Social Integration Confers Thermal Benefits in a Gregarious Primate," *Journal of Animal Ecology* 84, no. 3 (2015): 871–78.

9.  Tristen K. Inagaki et al., "A Pilot Study Examining Physical and Social Warmth: Higher (Non-febrile) Oral Temperature Is Associated with Greater Feelings of Social Connection," *PloS One* 11, no. 6 (2016): e0156873.

10. Pronobesh Banerjee, Promothesh Chatterjee, and Jayati Sinha, "Is It Light or Dark? Recalling Moral Behavior Changes Perception of Brightness," *Psychological Science* 23, no. 4 (2012): 407–9.

11. Dermot Lynott et al., "Replication of 'Experiencing Physical Warmth Promotes Interpersonal Warmth' by Williams and Bargh (2008)," *Social Psychology* (2014).

12. Colin F. Camerer et al., "Evaluating the Replicability of Social Science Experiments in *Nature* and *Science* between 2010 and 2015," *Nature Human Behaviour* 2, no. 9 (2018): 637.

13. Hans IJzerman et al., "The Human Penguin Project: Climate, Social Integration, and Core Body Temperature," *Collabra: Psychology* 4, no. 1 (2018).

14. Tal Yarkoni and Jacob Westfall, "Choosing Prediction over Explanation in Psychology: Lessons from Machine Learning," *Perspectives on Psychological Science* 12, no. 6 (2017): 1100–1122; Hans IJzerman et al., "What Predicts Stroop Performance? A Conditional Random Forest Approach," *SSRN Electronic Journal* (2016); Richard A. Klein et al., "Many Labs 2: Investigating Variation in Replicability across Samples and Settings," *Advances in Methods and Practices in Psychological Science* 1, no. 4 (2018): 443–90.

15. Everett Waters, David Corcoran, and Meltem Anafarta, "Attachment, Other Relationships, and the Theory That All Good Things Go Together," *Human Development* 48, no. 1–2 (2005): 80.

16. Hans IJzerman et al., "Socially Thermoregulated Thinking: How Past Experiences Matter in Thinking about Our Loved Ones," *Journal of Experimental Social Psychology* 79 (2018): 349–55.

17. Brian C. R. Bertram, "Vigilance and Group Size in Ostriches," *Animal Behaviour* 28, no. 1 (1980): 278–86.

18. Tsachi Ein-Dor, Mario Mikulincer, and Phillip R. Shaver, "Effective Reaction to Danger: Attachment Insecurities Predict Behavioral Reactions to an Experimentally Induced Threat above and beyond General Personality Traits," *Social Psychological and Personality Science* 2, no. 5 (2011): 467–73.

19. James A. Coan, Hillary S. Schaefer, and Richard J. Davidson, "Lending a Hand: Social Regulation of the Neural Response to Threat," *Psychological Science* 17, no. 12 (2006): 1032–39.

20. Tsachi Ein-Dor et al., "Sugarcoated Isolation: Evidence That Social Avoidance Is Linked to Higher Basal Glucose Levels and Higher Consumption of Glucose," *Frontiers in Psychology* 6 (2015): 492.

21. Rodrigo Clemente Vergara et al., "Development and Validation of the Social Thermoregulation and Risk Avoidance Questionnaire (STRAQ-1)," *International Review of Social Psychology* (in press).

22. V. Vuorenkoski et al., "The Effect of Cry Stimulus on the Temperature of the Lactating Breast of Primipara: A Thermographic Study," *Experientia* 25, no. 12 (1969): 1286–87.

23. Hans IJzerman et al., "A Theory of Social Thermoregulation in Human Primates," *Frontiers in Psychology* 6 (2015): 464.

24. Emily A. Butler and Ashley K. Randall, "Emotional Coregulation in Close Relationships," *Emotion Review* 5, no. 2 (2013): 202–10.

## CHAPTER 6
## Not by Hypothalamus Alone

1. Joel A. Allen, "The Influence of Physical Conditions in the Genesis of Species," *Radical Review* 1 (1877): 108–40.

2. Brett W. Carter and William G. Schucany, "Brown Adipose Tissue

in a Newborn," *Baylor University Medical Center Proceedings* 21, no. 3 (2008).

3. Bogusław Pawłwski, "Why Are Human Newborns So Big and Fat?," *Human Evolution* 13, no. 1 (1998): 65–72.

4. Frank E. Marino, "The Evolutionary Basis of Thermoregulation and Exercise Performance," in *Thermoregulation and Human Performance*, ed. Frank E. Marino (Basel, Switzerland: Karger Publishers, 2008), 53: 1–13.

5. Albert F. Bennett and John A. Ruben, "Endothermy and Activity in Vertebrates," *Science* 206, no. 4419 (1979): 649–54.

6. Dean Falk, "Brain Evolution in *Homo*: The 'Radiator' Theory," *Behavioral and Brain Sciences* 13, no. 2 (1990): 333–44.

7. Laura Tobias Gruss and Daniel Schmitt, "The Evolution of the Human Pelvis: Changing Adaptations to Bipedalism, Obstetrics and Thermoregulation," *Philosophical Transactions of the Royal Society B: Biological Sciences* 370, no. 1663 (2015): 20140063.

8. Peter E. Wheeler, "The Thermoregulatory Advantages of Hominid Bipedalism in Open Equatorial Environments: The Contribution of Increased Convective Heat Loss and Cutaneous Evaporative Cooling," *Journal of Human Evolution* 21, no. 2 (1991): 107–15.

9. Steven E. Churchill, "Bioenergetic Perspectives on Neanderthal Thermoregulatory and Activity Budgets," in *Neanderthals Revisited: New Approaches and Perspectives*, ed. Katerina Harvati and Terry Harrison (Dordrecht, Netherlands: Springer, 2006), 113–33.

10. Evelyn Satinoff, "Neural Organization and Evolution of Thermal Regulation in Mammals," *Science* 201, no. 4350 (1978): 16–22.

11. Michael L. Anderson, "Neural Reuse: A Fundamental Organizational Principle of the Brain," *Behavioral and Brain Sciences* 33, no. 4 (2010): 245–66.

12. K. A. Soudijn, G. J. M. Hutschemaekers, and F. J. R. van de Vijver, "Culture Conceptualisations," in *The Investigation of Culture: Current Issues in Cultural Psychology*, ed. F. J. R. van de Vijver and G. J. M. Hutschemaekers (Tilburg, Netherlands: Tilburg University Press, 1990), 19–39.

13. Harry C. Triandis, "Culture and Psychology: A History of the Study of Their Relationships," in *Handbook of Cultural Psychology*, ed. S. Kitayama and D. Cohen (New York: Guilford Press, 2007), 59–76.

14. Caroline Gilbert et al., "Huddling Behavior in Emperor Penguins: Dynamics of Huddling," *Physiology and Behavior* 88, no. 4–5 (2006): 479–88.

15. K. Bystrova et al., "Skin-to-Skin Contact May Reduce Negative Consequences of 'the Stress of Being Born': A Study on Temperature in Newborn Infants, Subjected to Different Ward Routines in St. Petersburg," *Acta Paediatrica* 92, no. 3 (2003): 320–26.

16. Ruth Feldman et al., "Skin-to-Skin Contact (Kangaroo Care) Promotes Self-Regulation in Premature Infants: Sleep-Wake Cyclicity, Arousal Modulation, and Sustained Exploration," *Developmental Psychology* 38, no. 2 (2002): 194.

17. A convenient compendium of recent research on animal tool use is Crickette M. Sanz, Josep Call, and Christophe Boesch, eds., *Tool Use in Animals: Cognition and Ecology* (Cambridge: Cambridge University Press, 2013).

18. Hans IJzerman and Francesco Foroni, "Not by Thoughts Alone: How Language Supersizes the Cognitive Toolkit," *Behavioral and Brain Sciences* 35, no. 4 (2012): 226.

19. Hans IJzerman and Gün R. Semin, "The Thermometer of Social Relations: Mapping Social Proximity on Temperature," *Psychological Science* 20, no. 10 (2009): 1214–20.

20. Andy Clark, *Supersizing the Mind: Embodiment, Action, and Cognitive Extension* (New York: Oxford University Press, 2008).

21. George Lakoff and Mark Johnson, *Metaphors We Live By* (Chicago: University of Chicago Press, 2008).

22. Zoltán Kövecses, *Metaphor in Culture: Universality and Variation* (Cambridge: Cambridge University Press, 2005).

23. Henrik Liljegren and Naseem Haider, "Facts, Feelings and Temperature Expressions in the Hindukush," in *The Linguistics of Temperature*, ed. Maria Koptjevskaja-Tamm (Amsterdam, Netherlands: John Benjamins, 2015), 440–70.

24. Poppy Siahaan, "Why Is It Not Cool? Temperature Terms in Indonesian," in *The Linguistics of Temperature*, ed. Maria Koptjevskaja-Tamm (Amsterdam, Netherlands: John Benjamins, 2015), 666–99.

25. Maria Koptjevskaja-Tamm, ed., *The Linguistics of Temperature* (Amsterdam, Netherlands: John Benjamins, 2015).

26. Peter J. Richerson and Robert Boyd, *Not by Genes Alone: How Cul-*

*ture Transformed Human Evolution* (Chicago: University of Chicago Press, 2008).

27.  Oliver G. Brooke, M. Harris, and Carmencita B. Salvosa, "The Response of Malnourished Babies to Cold," *Journal of Physiology* 233, no. 1 (1973): 75.

28.  Cara M. Wall-Scheffler, "Energetics, Locomotion, and Female Reproduction: Implications for Human Evolution," *Annual Review of Anthropology* 41 (2012): 71–85.

## CHAPTER 7
## Why You Should Sell Your House on a Colder Day

1.  Sue Williams, "Selling a House in Winter: How to Help Buyers Warm Up to Your Home," *Domain*, June 15, 2017, https://www.domain .com.au/news/homes-styled-and-built-for-warmth-shoot-ahead-in -sydneys-winter-property-market-20170608-gwndx7/.

2.  Larissa Dubecki, "Why There's a Hidden Advantage for Selling Your Home in Winter," *Domain*, June 9, 2017, https://www.domain.com .au/news/why-theres-a-hidden-advantage-for-selling-your-home-in -winter-20170609-gwdrkw/.

3.  "The Sweet Smell of Success: How Aroma Can Help You Sell Your Home," *Mountain Democrat*, July 19, 2011, https://www.mtdemocrat .com/business-real-estate/the-sweet-smell-of-success-how-aroma -can-help-you-sell-your-home/.

4.  Jiewen Hong and Yacheng Sun, "Warm It Up with Love: The Effect of Physical Coldness on Liking of Romance Movies," *Journal of Consumer Research* 39, no. 2 (2011): 293–306.

5.  Xinyue Zhou et al., "Heartwarming Memories: Nostalgia Maintains Physiological Comfort," *Emotion* 12, no. 4 (2012): 678.

6.  Lora E. Park and Jon K. Maner, "Does Self-Threat Promote Social Connection? The Role of Self-Esteem and Contingencies of Self-Worth," *Journal of Personality and Social Psychology* 96, no. 1 (2009): 203.

7.  Bram B. Van Acker et al., "Homelike Thermoregulation: How Physical Coldness Makes an Advertised House a Home," *Journal of Experimental Social Psychology* 67 (2016): 20–27.

8.  Hans IJzerman, Janneke A. Janssen, and James A. Coan, "Main-

taining Warm, Trusting Relationships with Brands: Increased Temperature Perceptions after Thinking of Communal Brands," *PloS One* 10, no. 4 (2015): e0125194.

9. Jan S. Slater, "Collecting Brand Loyalty: A Comparative Analysis of How Coca-Cola and Hallmark Use Collecting Behavior to Enhance Brand Loyalty," *ACR North American Advances* (2001).

10. Aaron C. Ahuvia, "I Love It!: Towards a Unifying Theory of Love across Diverse Love Objects (Abridged)" (Research Support, School of Business Administration, Working Paper No. 718, 1993).

11. Terence A. Shimp and Thomas J. Madden, "Consumer-Object Relations: A Conceptual Framework Based Analogously on Sternberg's Triangular Theory of Love," *ACR North American Advances* (1988).

12. Marsha L. Richins, "Measuring Emotions in the Consumption Experience," *Journal of Consumer Research* 24, no. 2 (1997): 127–46.

13. Joseph P. Simmons and Uri Simonsohn, "Power Posing: P-Curving the Evidence," *Psychological Science* (2017).

14. Bikhchandani Sushil and Sharma Sunil, "Herd Behavior in Financial Markets," *IMF Staff Papers* 48 (2001): 279–310.

15. Wayne D. Hoyer and D. J. MacInnis, *Consumer Behavior*, 3rd ed. (Boston: Houghton Mifflin, 2004).

16. Indeed, there are no *reliable* studies, but there are *some* studies on the topic, which is why I feel compelled to bring it up. In 2013, a report appeared on social thermoregulation and conformity by Huang and colleagues. I was strongly attracted to one of the studies in this report, on horse-race betting and temperature. When the mercury rises, the authors noted, people bet more frequently on the favorite horse. I was eager to include this study in my book as evidence of warmth increasing conformity, which seemed to me a striking manifestation of social thermoregulation. While I was writing, however, I received a tweet from a colleague, Quentin André, who pointed to some statistical errors in the Huang studies. He found, more specifically, that there were "granularity-related inconstancy of means" (or GRIM) errors. The GRIM test was developed by Nick Brown and James Heathers and is really cool, yet very simple, and you can even use it yourself to check accuracy in scientific papers you may read. In such papers, researchers typically report how many people they have tested (the sample size) and the average of the values in a condition

they test. Here's the cool part: for each sample size, it is possible to have only specific statistical means. Let's imagine that you test 28 participants and they all answer questions on a scale from 1 to 7. In your paper, you report an average of 5.19. This, however, cannot be correct. All responses were scored from 1 to 7. The scores must therefore fall between 28 and 159. The score that gives the average closest to 5.19 is 145 or 146. 145 divided by 28 is 5.17857, and 146 divided by 28 is 5.21429. There is no possibility to *average* a score of 5.19.

Quentin calculated the scores for the Huang study and found several GRIM errors. What caused these, we don't know at this point. It can be as simple as a rounding error. However, we also calculated the probability of the truth of the last study—my favorite: the real-life data on horse-race betting. We did this by essentially comparing the distribution of the "effect sizes" of other studies. The effect size tells us how strong an effect is. In this case, if the temperature changes by one degree, to what extent does conformity behavior change? We can then compare this to something wildly different, such as the height difference between men and women. If we then look at a distribution of effect sizes from other studies in our discipline, we can evaluate how probable the study in question is. In the instance of the horse-race betting study, we found that its reported results had only a one-in-a-billion chance of being true. Xun Irene Huang et al., "Warmth and Conformity: The Effects of Ambient Temperature on Product Preferences and Financial Decisions," *Journal of Consumer Psychology* 24, no. 2 (2014): 241–50.

17.   Pascal Bruno, Valentyna Melnyk, and Franziska Völckner, "Temperature and Emotions: Effects of Physical Temperature on Responses to Emotional Advertising," *International Journal of Research in Marketing* 34, no. 1 (2017): 302–20.

18.   Antonio Damasio and Hanna Damasio, "Minding the Body," *Daedalus* 135, no. 3 (2006): 15–22.

19.   Jeff D. Rotman, Seung Hwan Mark Lee, and Andrew W. Perkins, "The Warmth of Our Regrets: Managing Regret through Physiological Regulation and Consumption," *Journal of Consumer Psychology* 27, no. 2 (2017): 160–70.

20.   Yonat Zwebner, Leonard Lee, and Jacob Goldenberg, "The Tem-

perature Premium: Warm Temperatures Increase Product Valuation," *Journal of Consumer Psychology* 24, no. 2 (2014): 251–59.

21. Peter Kolb, Christine Gockel, and Lioba Werth, "The Effects of Temperature on Service Employees' Customer Orientation: An Experimental Approach," *Ergonomics* 55, no. 6 (2012): 621–35.

## CHAPTER 8
## From Depression to Cancer

1. Kay-U. Hanusch et al., "Whole-Body Hyperthermia for the Treatment of Major Depression: Associations with Thermoregulatory Cooling," *American Journal of Psychiatry* 170, no. 7 (2013): 802–4.

2. Philippa Howden-Chapman et al., "Tackling Cold Housing and Fuel Poverty in New Zealand: A Review of Policies, Research, and Health Impacts," *Energy Policy* 49 (2012): 134–42.

3. Harvey B. Simon, "Hyperthermia," *New England Journal of Medicine* 329, no. 7 (1993): 483–87.

4. Daniel F. Danzl and Robert S. Pozos, "Accidental Hypothermia," *New England Journal of Medicine* 331, no. 26 (1994): 1756–60.

5. Danzl and Pozos, "Accidental Hypothermia."

6. L. G. Pugh, "Accidental Hypothermia in Walkers, Climbers, and Campers: Report to the Medical Commission on Accident Prevention," *British Medical Journal* 1, no. 5480 (1966): 123.

7. Danzl and Pozos, "Accidental Hypothermia."

8. Hans IJzerman et al., "The Human Penguin Project: Climate, Social Integration, and Core Body Temperature," *Collabra: Psychology* 4, no. 1 (2018).

9. Takakazu Oka, Kae Oka, and Tetsuro Hori, "Mechanisms and Mediators of Psychological Stress-Induced Rise in Core Temperature," *Psychosomatic Medicine* 63, no. 3 (2001): 476–86.

10. Emails from Aaron Beck to Hans IJzerman, December 11, 2012: "I was very much interested in your piece in the New York Times on lowering of temperature following social exclusion. This fits very well with some of my notions of depression"; and December 17, 2012: "My interest in cooling effects of social exclusion comes from my

general interest in depression. I note that the excluded subject feels a coolness in the atmosphere and, indeed, does have a drop in temperature. I wonder whether the experience of social exclusion might trigger a more generalized cognitive bias which would be manifested by a general sense of being unacceptable, unlikable, unpopular, etc."

11. A. Wakeling and G. F. M. Russell, "Disturbances in the Regulation of Body Temperature in Anorexia Nervosa," *Psychological Medicine* 1, no. 1 (1970): 30–39.

12. A. W. Hetherington and S. W. Ranson, "Hypothalamic Lesions and Adiposity in the Rat," *Anatomical Record* 78, no. 2 (1940): 149–72.

13. Bal K. Anand and John R. Brobeck, "Hypothalamic Control of Food Intake in Rats and Cats," *Yale Journal of Biology and Medicine* 24, no. 2 (1951): 123.

14. Bengt Andersson and Börje Larsson, "Influence of Local Temperature Changes in the Preoptic Area and Rostral Hypothalamus on the Regulation of Food and Water Intake," *Acta Physiologica Scandinavica* 52, no. 1 (1961): 75–89.

15. C. L. Hamilton and John R. Brobeck, "Food Intake and Temperature Regulation in Rats with Rostral Hypothalamic Lesions," *American Journal of Physiology—Legacy Content* 207, no. 2 (1964): 291–97.

16. C. J. De Vile et al., "Obesity in Childhood Craniopharyngioma: Relation to Post-operative Hypothalamic Damage Shown by Magnetic Resonance Imaging," *Journal of Clinical Endocrinology and Metabolism* 81, no. 7 (1996): 2734–37.

17. Andrew Wit and S. C. Wang, "Temperature-Sensitive Neurons in Preoptic-Anterior Hypothalamic Region: Actions of Pyrogen and Acetylsalicylate," *American Journal of Physiology—Legacy Content* 215, no. 5 (1968): 1160–69.

18. Charles L. Raison et al., "Somatic Influences on Subjective Well-Being and Affective Disorders: The Convergence of Thermosensory and Central Serotonergic Systems," *Frontiers in Psychology* 5 (2015): 1580.

19. Nicholas G. Ward, Hans O. Doerr, and Michael C. Storrie, "Skin Conductance: A Potentially Sensitive Test for Depression," *Psychiatry Research* 10, no. 4 (1983): 295–302.

20. Irina A. Strigo, Alan N. Simmons, and Scott C. Matthews, "Increased Affective Bias Revealed Using Experimental Graded Heat Stimuli in

Young Depressed Adults: Evidence of 'Emotional Allodynia,'" *Psychosomatic Medicine* 70, no. 3 (2008): 338.

21. Alexander Ushinsky et al., "Further Evidence of Emotional Allodynia in Unmedicated Young Adults with Major Depressive Disorder," *PloS One* 8, no. 11 (2013): e80507.

22. L.-H. Thorell, "Valid Electrodermal Hyporeactivity for Depressive Suicidal Propensity Offers Links to Cognitive Theory," *Acta Psychiatrica Scandinavica* 119, no. 5 (2009): 338–49.

23. Matthew W. Hale et al., "Evidence for In Vivo Thermosensitivity of Serotonergic Neurons in the Rat Dorsal Raphe Nucleus and Raphe Pallidus Nucleus Implicated in Thermoregulatory Cooling," *Experimental Neurology* 227, no. 2 (2011): 264–78.

24. These records are summarized in "Wim Hof," Wikipedia, https://en.wikipedia.org/wiki/Wim_Hof.

25. Hof cites the intended benefits on his site: https://www.wimhofmethod.com/benefits.

26. Otto Muzik, Kaice T. Reilly, and Vaibhav A. Diwadkar, "'Brain over Body': A Study on the Willful Regulation of Autonomic Function during Cold Exposure," *NeuroImage* 172 (2018): 632–41.

27. Wouter van Marken Lichtenbelt, "Who Is the Iceman?," *Temperature* 4, no. 3 (2017): 202.

28. Mark J. W. Hanssen et al., "Short-Term Cold Acclimation Improves Insulin Sensitivity in Patients with Type 2 Diabetes Mellitus," *Nature Medicine* 21, no. 8 (2015): 863.

29. Hale et al., "Evidence for In Vivo Thermosensitivity," 264–78.

30. Brant P. Hasler et al., "Phase Relationships between Core Body Temperature, Melatonin, and Sleep Are Associated with Depression Severity: Further Evidence for Circadian Misalignment in Nonseasonal Depression," *Psychiatry Research* 178, no. 1 (2010): 205–7.

31. Gregory M. Brown, "Light, Melatonin and the Sleep-Wake Cycle," *Journal of Psychiatry and Neuroscience* 19, no. 5 (1994): 345.

32. National Cancer Institute, "Hyperthermia in Cancer Treatment," https://bit.ly/35vZ78H.

33. A. Merla and G. L. Romani, "Functional Infrared Imaging in Medicine: A Quantitative Diagnostic Approach," *2006 International Conference of the IEEE Engineering in Medicine and Biology Society* (IEEE, 2006).

34. Thorsten M. Buzug et al., "Functional Infrared Imaging for Skin-Cancer Screening," *2006 International Conference of the IEEE Engineering in Medicine and Biology Society* (IEEE, 2006).

35. Maria Tsoli et al., "Activation of Thermogenesis in Brown Adipose Tissue and Dysregulated Lipid Metabolism Associated with Cancer Cachexia in Mice," *Cancer Research* 72, no. 17 (2012): 4372–82.

36. Rajan Singh et al., "Increased Expression of Beige/Brown Adipose Markers from Host and Breast Cancer Cells Influence Xenograft Formation in Mice," *Molecular Cancer Research* 14, no. 1 (2016): 78–92.

37. Takaaki Fujii et al., "Implication of Atypical Supraclavicular F18-Fluorodeoxyglucose Uptake in Patients with Breast Cancer: Association between Brown Adipose Tissue and Breast Cancer," *Oncology Letters* 14, no. 6 (2017): 7025–30; Miriam A. Bredella et al., "Positive Effects of Brown Adipose Tissue on Femoral Bone Structure," *Bone* 58 (2014): 55–58.

38. Cacioppo has written a lot about loneliness, but one particularly interesting talk was his TED Talk in Des Moines: https://www.youtube.com/watch?v=_0hxl03JoA0.

39. Julianne Holt-Lunstad, Timothy B. Smith, and J. Bradley Layton, "Social Relationships and Mortality Risk: A Meta-Analytic Review," *PLoS Medicine* 7, no. 7 (2010): e1000316.

40. Julianne Holt-Lunstad et al., "Loneliness and Social Isolation as Risk Factors for Mortality: A Meta-Analytic Review," *Perspectives on Psychological Science* 10, no. 2 (2015): 227–37.

41. Steven D. Targum and Norman Rosenthal, "Seasonal Affective Disorder," *Psychiatry (Edgmont)* 5, no. 5 (2008): 31.

42. Marcus J. H. Huibers et al., "Does the Weather Make Us Sad? Meteorological Determinants of Mood and Depression in the General Population," *Psychiatry Research* 180, no. 2–3 (2010): 143–46.

43. WHO International Programme on Chemical Safety, "Biomarkers in Risk Assessment: Validity and Validation," *Environmental Health Criteria* 222 (2001), http://www.inchem.org/documents/ehc/ehc/ehc222.htm.

44. Sami Timimi, "No More Psychiatric Labels: Why Formal Psychiatric Diagnostic Systems Should Be Abolished," *International Journal of Clinical and Health Psychology* 14, no. 3 (2014): 208–15.

45. Eiko I. Fried, "The 52 Symptoms of Major Depression: Lack of Content Overlap among Seven Common Depression Scales," *Journal of Affective Disorders* 208 (2017): 191–97.

## CHAPTER 9
## The Happy Costa Ricans

1. Christopher Columbus, "Third Voyage," in J. M. Cohen, trans., *Christopher Columbus: The Four Voyages* (London: Penguin, 1969), 221, 219.

2. Samuli Helama, Jari Holopainen, and Timo Partonen, "Temperature-Associated Suicide Mortality: Contrasting Roles of Climatic Warming and the Suicide Prevention Program in Finland," *Environmental Health and Preventive Medicine* 18, no. 5 (2013): 349; Reija Ruuhela et al., "Climate Impact on Suicide Rates in Finland from 1971 to 2003," *International Journal of Biometeorology* 53, no. 2 (2009): 167.

3. Christian Bjørnskov, Axel Dreher, and Justina A. V. Fischer, "Cross-Country Determinants of Life Satisfaction: Exploring Different Determinants across Groups in Society," *Social Choice and Welfare* 30, no. 1 (2008): 119–73.

4. The history of the World Values Survey can be read on the World Values Survey Association web page: http://www.worldvaluessurvey.org/WVSContents.jsp?CMSID=History.

5. National Institute of Mental Health, "Seasonal Affective Disorder," last revised March 2016, https://www.nimh.nih.gov/health/topics/seasonal-affective-disorder/index.shtml.

6. Michelle L. Taylor et al., "Temperature Can Shape a Cline in Polyandry, but Only Genetic Variation Can Sustain It over Time," *Behavioral Ecology* 27, no. 2 (2016): 462–69, https://www.ncbi.nlm.nih.gov/pmc/articles/PMC4797379/.

7. Peter Paul A. Mersch et al., "Seasonal Affective Disorder and Latitude: A Review of the Literature," *Journal of Affective Disorders* 53, no. 1 (1999): 35–48.

8. Leora N. Rosen et al., "Prevalence of Seasonal Affective Disorder at Four Latitudes," *Psychiatry Research* 31, no. 2 (1990): 131–44.

9.  J. Henrich, S. J. Heine, and A. Norenzayan, "The Weirdest People in the World?," *Behavioral and Brain Sciences* 33, no. 2–3 (2010): 61–83.

10. Rachid Laajaj et al., "Challenges to Capture the Big Five Personality Traits in Non-WEIRD Populations," *Science Advances* 5, no. 7 (2019): eaaw5226.

11. Emorie D. Beck, David M. Condon, and Joshua J. Jackson, "Interindividual Age Differences in Personality Structure," PsyArxiv (July 19, 2019), https://psyarxiv.com/857ev/.

12. J. K. Flake and E. I. Fried, "Measurement Schmeasurement: Questionable Measurement Practices and How to Avoid Them," PsyArXiv (January 17, 2019), doi:10.31234/osf.io/hs7wm.

13. Eiko I. Fried, "The 52 Symptoms of Major Depression: Lack of Content Overlap among Seven Common Depression Scales," *Journal of Affective Disorders* 208 (2017): 191–97.

14. Eiko I. Fried and Randolph M. Nesse, "Depression Sum-Scores Don't Add Up: Why Analyzing Specific Depression Symptoms Is Essential," *BMC Medicine* 13, no. 1 (2015): 72.

15. Paul A. M. Van Lange, Maria I. Rinderu, and Brad J. Bushman, "Aggression and Violence around the World: A Model of CLimate, Aggression, and Self-Control in Humans (CLASH)," *Behavioral and Brain Sciences* 40 (2017).

16. Craig A. Anderson, "Temperature and Aggression: Ubiquitous Effects of Heat on Occurrence of Human Violence," *Psychological Bulletin* 106, no. 1 (1989): 74.

17. Hans IJzerman et al., "Does Distance from the Equator Predict Self-Control? Lessons from the Human Penguin Project," *Behavioral and Brain Sciences* 40 (2017): e86.

18. Jared Diamond, *Guns, Germs, and Steel: The Fates of Human Societies* (New York: W. W. Norton, 1997).

# Index

Note: Page numbers followed by "t" indicate tables.

4/22
3